ENGINEERING FOR YOU

A CAREER GUIDE

ENGINEERING FOR YOU

A CAREER GUIDE

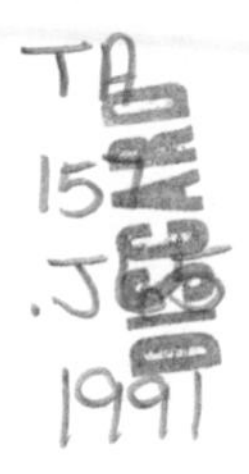

JOHN TAYLOR JONES

IOWA STATE UNIVERSITY PRESS / AMES

TO THE MEMORY OF

GEORGE R. TOWN (1905-1978)

Dean, College of Engineering, Iowa State University, 1959–1970

John Taylor Jones is vice president of research and development for Lenox China/Crystal, Absecon, New Jersey.

Manufactured in the United States of America
⊗ This book is printed on acid-free paper.

First edition, 1991

Library of Congress Cataloging-in-Publication Data

Jones, John Taylor
Engineering for you: a career guide / John Taylor Jones. — 1st ed.
p. cm.
Includes index.
ISBN 0-8138-0603-8
1. Engineering — Vocational guidance. I. Title
TA157.J65 1991
620′.0023 — dc20 90-43467

CONTENTS

PART THREE: SHOULD YOU BE AN ENGINEER?

PREFACE

At no time in history has there been a greater need to have young men and women enter the engineering profession. Problems that are pressing hard on the nerves of society such as international competition, energy availability, pollution, housing, food production and processing, material shortages, and transportation require in the main engineering solutions. The public knows this and expects engineers to play their role to solve such problems permanently and expeditiously.

There are no simple solutions to the major crises facing the world. Solutions are complex, requiring the knowledge and skills and cooperation of engineers, scientists, sociologists, and governments. (Engineers from a number of different engineering fields must combine their knowledge to solve the technical aspects of societal problems.) Engineers of the future will need a broadly based education, yet they will also be required to specialize more if they are to contribute to teams of problem solvers from a variety of disciplines.

The future of engineering is ensured by the fact that society continues to create needs and problems for engineers to alleviate. There is room for women and men with broad interest in the arts and sciences, for inventors, for technologists — and for you if you are looking for a challenge and have the interest and ability. This book was written to help you make the decision — is engineering for you?

Acknowledgments

I thank the following Iowa State University staff members: Dean Emeritus George R. Town, deceased, who "championed" this book and to whom this book is dedicated, and Associate Dean Paul Morgan, David Martin, Tom McGee, and Orville Hunter for reading the manuscript and recommending a number of useful additions and changes. I also thank Theodore Okiishi, Jack Mickle, Robert Brodsky, Arvid Eide, Maurice Larson, Edgar Collins, Howard Johnson, Donald McKeown, Richard Handy, Carl Ekberg, Charles Cowan, Richard Squires, and Richard Danofsky for reviewing the sections on the various engineering fields and adding useful information. David R. Reyes-Guerra, former executive director of the Engineers' Council for Professional Development (ECPD) and currently executive director of the Accreditation Board for Engineering and Technology, Inc., (ABET) was particularly helpful.

The following also supplied much helpful information: Leslie-Anne Boss, marketing coordinator, Publication Group, Society of Automotive Engineers; Society of Petroleum Engineers; Society of Women Engineers; Barbara A. Boyer, coordinator, Student Member Activities, American Institute of Chemical Engineers; Greg Fischer, director of technical services, American Ceramic Society; Sharon E. Connelly, education coordinator, Society of Mining Engineers; Pamela A. Cole, Member Services, American Society of Agricultural Engineers; William R. Anderson, administrator, Professional Program, Institute of Electrical and Electronics Engineers; and Amanda F. Guthridge, university chapter coordinator, Institute of Industrial Engineers. And thanks to my son, Jimmy, who has been looking over my shoulder and helping me place the photographs that were generously supplied by the U.S. Department of the Interior, NASA, McDonnell Douglas, The Norton Company, Phillips Petroleum Company, Chrysler Corporation, Ford Motor Company, and General Motors, Inc.

I wish to thank Mary Brown, Barbara Briese, and Fortune Scaraglino, who prepared the manuscript. Finally, I thank my wife, Pat, for suggesting that interviews with working engineers be added to the manuscript.

INTRODUCTION

A person's life span is short when compared with the age of our planet, Earth. The very existence of humankind seems almost trivial from a cold, scientific point of view. What influence can we have on the universe? Can we change the orbit of our planet? Can we keep the sun from dying? Can we overcome our own death or travel through interstellar space to find a new Earth for future generations? I am sure many would say that we can have little influence on the universe, that the answers to those questions are all no.

Fortunately, there are those who believe in a great destiny for humankind, who believe there will be solutions to problems that now seem unsolvable. They enjoy working hard at tasks that will benefit their society. The world needs these valuable people. We need them in engineering.

There are two difficult decisions to be made by students when considering engineering as a career. The first is to decide whether or not engineering fits in with their abilities and life's goals. The second is to choose fields of engineering consistent with their particular career objectives. To help you make these important decisions this book is divided into three parts. The first illustrates the engineer's place in society, the second describes the fields of engineering, and the third presents criteria for deciding which professional engineering roles you might enjoy.

PART ONE

THE ENGINEER'S PLACE IN SOCIETY

A generally accepted definition of engineering, originated by the Engineers' Council for Professional Development and now used by the Accreditation Board for Engineering and Technology and by the American Association of Engineering Societies, is

> Engineering is the profession in which a knowledge of the mathematical and natural sciences gained by study, experience, and practice is applied with judgment to develop ways to utilize, economically, the materials and forces of nature for the benefit of mankind.

The first part of this book on the engineer's place in society will expand on this definition. Chapter 1 explores the importance of science and mathematics to engineers. Chapter 2 explains that design is the essence of all engineering. Chapter 3 describes the engineer's professionalism, and Chapter 4 explains the engineer's function in industry.

1 SCIENCE AND ENGINEERING

A young person of high school age has usually developed a set of traits, abilities, and interests that, if analyzed carefully by a competent guidance counselor (or other qualified professional person), suggest a general area of vocational preference. However, because there are such varied personality types working in both science and engineering and since the basic starting credentials are the same for both of these major professional areas, the final decision as to which profession an individual should enter is difficult. To make the proper choice, a young person must find the answers to two questions: What is science? What is engineering?

A high school student is usually more familiar with science than with engineering. Many high school teachers majored in science in college and teach the methods and philosophy of science in the classroom and the laboratory. The student knows that scientists gather knowledge by observation and experimentation and categorize it under various fields of knowledge. Hypotheses are made as to why nature behaves as it does, and if a large amount of data fit a hypothesis it may be expanded into a general theory and perhaps a law.

Although engineers use the findings of science, they are not scientists. In fact, engineers frequently make decisions in areas where scientific knowledge is completely void.

Even though engineers are not classical scientists, their education in science and mathematics is extremely intense. It is not unusual for a chemical engineer to take more chemistry courses than a chemistry major or for a computer or electrical engineer to have more mathematics than a mathematics major. Because they apply their scientific knowledge, they learn where deficiencies in the basic science exist. Many of our important scientific principles and theories were discovered by engineers.

The federal government considers engineers to be part of our scientific work force. In fact, they make up the largest group in the country with a science and mathematics basic education.

Engineering is the effort of humankind to control and utilize nature for the benefit of society. All engineers are taught how to solve problems and how to design. They create devices, provide manufactured products, build structures, devise systems and services, and perform many functions that benefit humankind.

You can determine your inclination toward engineering if you study your answers to questions such as the following.

Which would you rather do:

1. (a) Design a large concrete dam, or
   (b) Study the chemical composition of cement?
2. (a) Design a television set, or
   (b) Study the absorption of light by liquids?
3. (a) Design a plant to make plastic sheet, or
   (b) Conduct experiments on polymerization material in the laboratory?
4. (a) Design a space vehicle to go to Jupiter, or
   (b) Analyze the data concerning the chemical composition of Jupiter returned to earth by radio?
5. (a) Manage an automobile factory, or
   (b) Conduct chemical analysis in a chemical laboratory?

If your answers to the questions were predominantly (a), then you are already inclined toward engineering. If you had a difficult time deciding between the various choices, you still are inclined to engineering. (It is true that many engineers sometimes do scientific research, and they are recognized as scientists as well as engineers. The transition from scientist to engineer is not as common.)

Despite the fact that engineering is a different activity from

1.1 Kennedy Space Center, Florida. A McDonnell Douglas engineer works on a modification to Atlantis's left-hand Orbital Maneuver System (OMS) pod (LP04). The modification involves installation of a new system that will make it easier for technicians to check for the presence of telltale propellant vapors that would indicate leakage in valve components. The work is being performed at the Space Center's Hyperbolic Maintenance Facility to ready the pod for use on Atlantis's next mission. For flight, the OMS pods are bolted to the aft fuselage of the shuttle orbiter and contain the engines and thrusters used to maneuver the spaceship in orbit. (Photo courtesy of NASA)

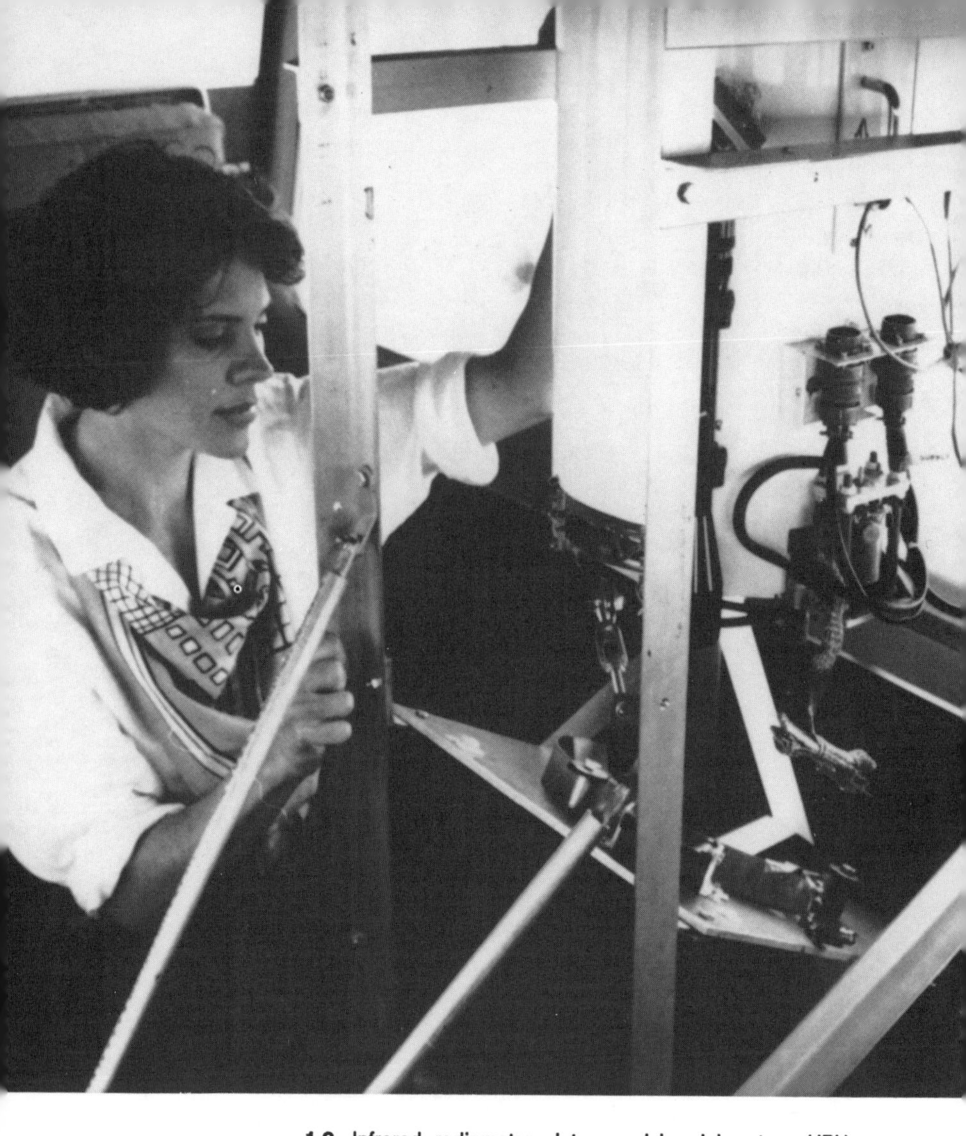

1.2. Infrared radiometer. Jet propulsion laboratory (JPL) scientists have developed this highly sensitive device to measure thermal radiation at the ocean's surface. They have test flown the instrument over the San Pedro harbor aboard the Goodyear blimp "Columbia." An instrument-ladened buoy deployed from a boat below the blimp took similar readings for comparative purposes. Knowing how much thermal radiation is present is important in tracking global weather patterns. Currently weather satellites, equipped with less sophisticated radiometers, are used to measure this radiation. The goal of the JPL effort is to evaluate the reliability of this satellite data. (Photo courtesy of NASA)

science, there is a distinct relationship between the two activities. A historical example used by Dean George R. Town (1905–1978), Iowa State University, might be useful at this point. A great English mathematician by the name of James Clark Maxwell (1831–1879) first predicted that electromagnetic waves should exist. A German physicist, Heinrich Hertz, later proved experimentally in the laboratory (circa 1879) that Maxwell's premise was correct. The entire system of radio and television broadcasting was then developed by engineers. It is obvious from this example that engineers *apply* scientific knowledge more than they search for it.

You can see that engineers need a good understanding of scientific concepts if they are going to be able to apply them. Mathematics and a good knowledge of either chemistry or physics (preferably both) are required.

Scientists and engineers may work on the same projects. The space program is a good example — engineers design the entire system (the spacecraft, the telecommunications, the navigation system, the launching apparatus, the computer, the software), and scientists evaluate the data returned from various instruments by radio.

We can see how scientists and engineers work together in industry if we answer the question, How does a new product evolve? A simple case would be as follows: A scientists discovers a new compound in the laboratory, determines its molecular weight and structure, and gives it a name, say, "Compound X." The scientist predicts other properties it may have but then returns to the laboratory to search for "Compound Y." The laboratory director decides to study "Compound X" further and determines that it is a plastic material that may be useful to society, that its properties are different from available materials, and that the cost of manufacture may be relatively low. The information regarding the synthesis and properties of the material is given to the head *development engineer.* This person studies the information and directs subordinate development engineers to set up a small operation that will yield enough material to determine the engineering properties and to supply sales engineers (or other salespeople) with samples for potential customer evaluation. If enough interest is shown, a pilot plant (a small production operation) large enough to handle initial customer demand is constructed. If the demand grows, a full-size manufacturing operation is constructed using information determined from the pilot operation.

From the last example, you might raise the question, "Where does the engineer obtain the information to design a pilot plant or a full-scale manufacturing operation?" The answer is that this is the type of information engineers have that most other people don't have. It is a treasure trove of knowledge and skills developed over the centuries. The engineer uses the principles of engineering design to create new things. In the next chapter, we will look at the design process and its importance to engineering.

2 THE ENGINEER AS A DESIGNER

The true essence of engineering is design. Design is the heart of all creative thinking whether it is in music, art, mathematics, or engineering. Design is a process—the most important part of which is thinking. It is a natural process that all humans use. In the typical case, a person is confused with a problem. He or she ponders the problem—perhaps searching for more information to help solve it. Eventually, ideas cross the mind that may lead to a solution. Perhaps after some time a stupendous idea comes (classically, cartoonists show this idea as a lighted bulb over the person's head), which the individual can develop until the problem is solved.

The steps in the design process have been listed by many educators, writers, artists, engineers, and others. I am sure that you could list them yourself if you thought about it, and your list would probably be as good as anyone's. The actual words you would list would probably relate to the particular design activity you were engaged in. Alger and Hays, in their book (a part of the Prentice-Hall Series in Engineering Design) *Creative Synthesis in Design* (1964), used the following terminology:

Recognizing—Does a problem really exist? What is the exact problem? Should it be restated?

Specifying—What is actually needed? For an aircraft you might specify speed, size, special features, and fuel consumption.

Proposing solutions—With all possible creativity, possible solutions, no matter how strange, would be listed.

Evaluating alternatives — Which solutions are best?

Deciding on a solution — Of the best solutions, which one should be selected?

Like all designers, engineers need to develop creativity. How creative a person is depends not only on education but also on everyday life from the time of birth. One tactic to improve creativity is to brainstorm. In brainstorming, a group of individuals having different backgrounds get together and dream up ideas or possible problem solutions. Many of the best ideas may come from the more creative minds in the group. Brainstorming can help a person improve creativity. It also permits the interaction of people from different vocational disciplines.

Knowledge of a broad nature is important in creativity. The creative person crosses the lines between seemingly unrelated fields. Let me give you a simplified idea of how that works. I want you to read the next sentence and then do what it says (mentally or using a piece of scratch paper) without looking at the following paragraph. **Design a mousetrap.**

I assume now that you have designed a mousetrap at least mentally. You probably went through a process like this: He wants me to design a mousetrap. Sounds silly—a mousetrap is a small board with a spring-loaded killing wire (mental image). I think he wants something different from that. Let's see. I could use poison. Would that be a trap? Oh, I know, I will ________ them!

If you said mentally, "I will ________ them," you probably went into a new field (different from spring technology) for your solution. You also redefined the problem to *kill mice* rather than *build mousetrap.* For example, if you said, "I will electrocute them," you went into the electrical field for your solution. Restating the problem and crossing the fields-of-learning lines are important parts of creative design.

Engineers need much more than creativity to do their work. In the case of the electric mousetrap, an engineer would need specific knowledge about electrical circuits, shock hazards, and the laws governing electrical appliances.

2.1. Chrysler corporations's designers and engineers use the computer-aided design (CAD) system to create bodies of vehicles, including the underbody and panels, steering geometry, suspension, and other systems. Shown here on a CAD terminal at Chrysler is a Chrysler LeBaron GTS/Dodge Lancer that can be fitted with components, passengers, and cargo. (Photo courtesy of Chrysler Corporation)

Because engineering design is frequently on the fringe of the unknown, there may be repercussions detrimental to particular individuals or to society in general. Consequently, the engineer needs to be a prophet to protect the public. If she or he is a false prophet, society suffers and the engineering profession suffers along with it.

Partly for this reason, and others, engineers have joined into professional societies to control the application of engineering. Only engineers who meet certain stringent requirements and pass state examinations are allowed to supervise public projects and to take responsibility for structures and products in which the safety of the public, or of the user, is involved. They are called registered professional engineers. Like all professions, engineering tries to maintain high standards of performance, and of ethics, and thereby gain the respect of the public.

3 THE ENGINEER AS A PROFESSIONAL

Engineering is a profession because in engineering knowledge gained through high-level, specialized training is used responsibly and ethically in the service of humankind. The term *profession* does not hold exactly the same meaning for engineering as it does for medicine, music, or the arts, in which many professionals are self-employed. Most engineers, with the exception of consulting engineers who have their own engineering businesses, are employed in industrial or governmental organizations. Many consulting engineers and architectural engineers have a direct personal relationship with their clients, which is an essential element of professionalism.

Most professions have a code of ethics or responsibility. Engineering is no exception. An example provided by the American Association of Engineering Societies is given here:

FAITH OF THE ENGINEER

I AM AN ENGINEER. In my profession I take deep pride, but without vainglory; to it I owe solemn obligations that I am eager to fulfill.

As an Engineer, I will participate in none but honest enterprise. To him that has engaged my services, as employer or client, I will give the utmost of performance and fidelity.

> When needed, my skill and knowledge shall be given without reservation for the public good. From special capacity springs the obligation to use it well in the service of humanity; and I accept the challenge that this implies.
>
> Jealous of the high repute of my calling, I will strive to protect the interests and the good name of any engineer that I know to be deserving; but I will not shrink, should duty dictate, from disclosing the truth regarding anyone that, by unscrupulous act, has shown himself unworthy of the profession.
>
> Since the Age of Stone, human progress has been conditioned by the genius of my professional forbears. By them have been rendered usable to mankind Nature's vast resources of material and energy. By them have been vitalized and turned to practical account the principles of science and the revelations of technology. Except for this heritage of accumulated experience, my efforts would be feeble. I dedicate myself to the dissemination of engineering knowledge, and, especially to the instruction of younger members of my profession in all its arts and traditions.
>
> To my fellows I pledge, in the same full measure I ask of them, integrity and fair dealing, tolerance and respect, and devotion to the standards and the dignity of our profession; with the consciousness, always, that our special expertness carries with it the obligation to serve humanity with complete sincerity.

An engineer's responsibilities can be summarized under the headings of the employer, the public, and the engineer. Since knowledge is money to an employer, the engineer should not spread trade secrets through the specific industry he or she is working in. A day's work should be given for a day's pay, and the engineer should not spend time complaining or bickering on or about the job. If an engineer doesn't like a particular job and the problems cannot be rectified by honest communication with the employer, he or she should go somewhere else. If an engineer does not have the competence to do a particular job right, that person should forewarn the employer that assistance will be needed.

Engineers deal with the public in a number of ways. It is not difficult to see that engineers must take every precaution to protect the public from serious accidents and loss of money. If a tornado destroys a city, taking lives along its path, people understand that it could not be prevented. But if a dam collapses, drowning hundreds, the responsibility immediately is placed on the agency that constructed it. If the dam was constructed under the directions of engineers, they must take the blame for the failure unless the circumstances of failure are unusual indeed. Engineering firms should not take undue risks to permit lower bidding on construction, and workers must be informed of important safety considerations during construction and operation of structures and equipment.

Engineers often work with safety factors. For example, a safety factor of two would mean that a particular structural component would withstand twice the expected stresses on the component. However, the best safety factor is a competent, responsible engineer. It should be a daily part of every engineer's life to try to improve personal technical and managerial abilities.

One of the most exciting challenges to engineers is the competitive world of business as discussed in the next chapter. You will see that the engineering profession is for competent and responsible people.

4 THE ENGINEER IN INDUSTRY

Most engineers work in industry or governmental service. Industry recruits heavily on college campuses, and the engineer's first job is usually in industry. Civil engineering is an exception. A substantial number of civil engineers are employed in public works or with private consulting firms.

Industry implies the manufacture of products. Whether a product is a child's toy or a spaceship, the principles of manufacturing are the same. A simplified cycle is as follows:

1. A potential market for a certain product is verified.
2. The product is designed, constructed, and tested for efficiency, reliability, and cost.
3. The product is marketed on a pilot scale to determine pricing and customer satisfaction.
4. The total market is estimated.
5. The production process is designed and tested (based on pilot plant data).
6. Production begins, and testing continues to ensure quality maintenance.
7. Sales promotions are modified and tested to increase sales and profits.
8. The product is modified based on field test results to maintain consumer interest and acceptance.
9. When the product becomes obsolete, the production process

is terminated or modified for new products.

10. Surplus production is discounted and sold as salvage.
11. The search for new product ideas continues.

This process requires individuals with various temperaments, knowledge, and abilities. Interestingly enough, engineers may be engaged in any of the steps in the cycle. A supervising engineer or manager may have jurisdiction in several or all of the above steps.

Beginning engineers have several ways to start their careers in industry. They may start at the beginning of a product cycle in design or marketing. They may work as quality assurance engineers or in production engineering. Or they may even start in sales engineering. Individual entry-level positions depend on the desires and abilities of the graduating engineering senior.

No matter where an engineer starts, that person will soon be supervising others. There may be draftspersons, technicians, skilled workers, and/or laborers to start. Even a beginning design engineer may have several technicians if creative and able to supervise others.

The line of promotion usually looks something like this:

1. Engineer trainee or junior engineer
2. Engineer
3. Senior engineer
4. Engineering or production supervisor
5. Engineering manager or production superintendent
6. V.P. of engineering or V.P. of production
7. President
8. Chairman of the board of directors (about one-half of the one hundred largest corporations in the United States have engineers as chairman of the board).

The smaller companies often have fewer job descriptions. An employee may start as an engineer. The next steps may be department head, production superintendent, and then plant manager. The engineer advancing in administration must learn managerial skills while continuing to rely on engineering skills. Most companies have educational and training programs to help engineers learn these new skills. The engineering student who wishes to go into management should take business administration courses while still in college.

4.1. Manufacturing engineers on the line. (Photo courtesy of General Motors, Inc.)

Between top management and the sales force, many engineers find themselves in production, quality assurance, and development work. A production engineer carries more responsibilities than the average quality assurance or development engineer and often receives a higher salary. Production engineers need a good set of business skills and may want to supplement their engineering education after graduation.

Production engineering supervisors spend much time scheduling production; hiring, firing, and arbitrating with employees and union officers; and putting out fires (correcting production problems). Production problems have first priority in most companies, and the production engineering supervisor often defines problems to quality assurance and development people and nonsupervisory production engineers. These people in turn solve the problems. Production engineers frequently suggest and implement ways to improve production processes.

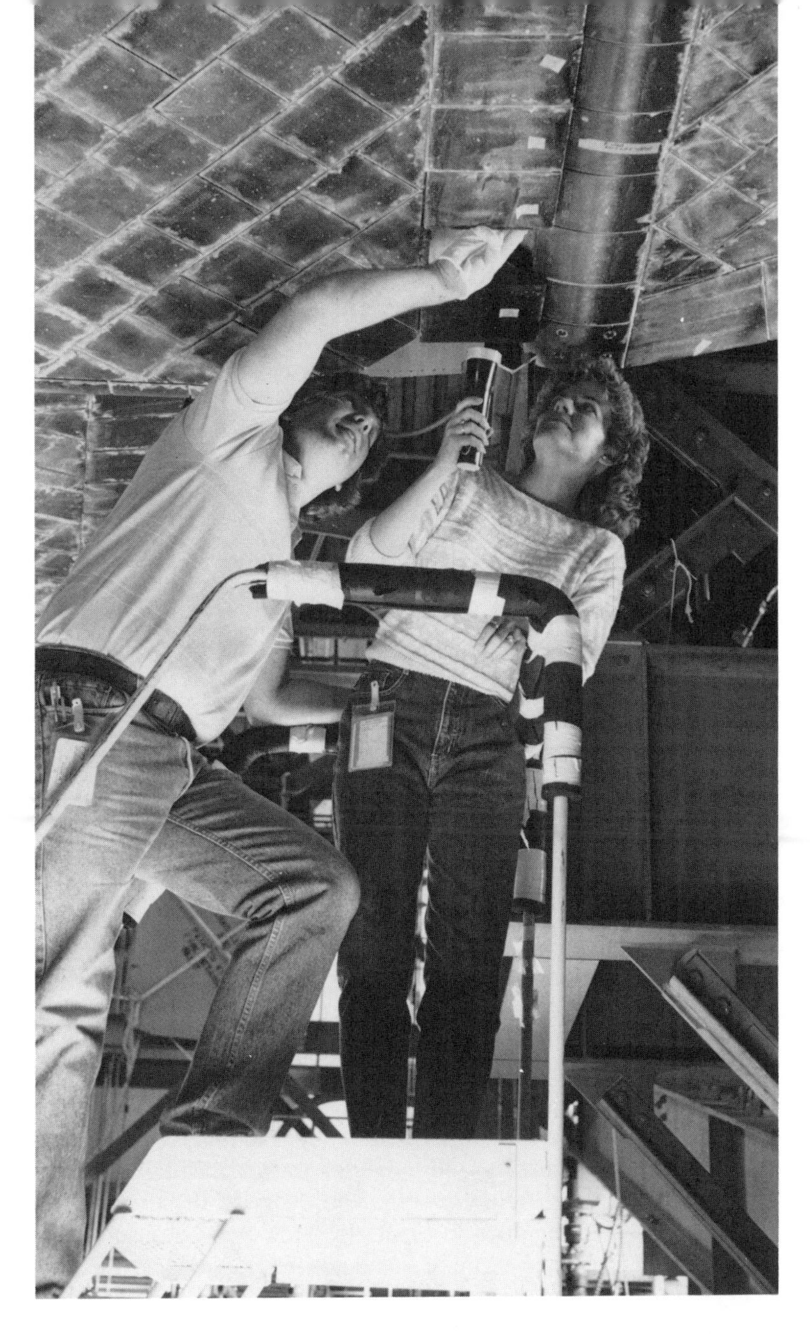

4.2. Kennedy Space Center, Florida. Workers examine the area where the body flap meets the aft fuselage as structural inspections continue on the Discover in High Bay 1 of the Orbiter Processing Facility. (Photo courtesy of NASA)

4.3. Particle paths on modified F16A, on IRIS workstation. Powerful computational tools permit development engineers to predict aerodynamic parameters before modeling for actual wind tunnel experiments. (Photo courtesy of NASA)

Quality assurance engineers maintain control of the production process and search for sources of flaws in the final product. Some companies use quality assurance as a training ground for their production engineers because it is a good starting point from which to learn the overall operation of a company. Quality assurance engineers should know statistics and be familiar with various measurement and quality and process control techniques. Each engineering discipline requires some knowledge of these areas, and industrial engineering does so especially. However, because quality assurance requires even more of such knowledge, young engineering graduates in quality assurance may want to supplement their education.

The *development engineer* (as mentioned in Chapter 1) is an intermediate between research and production. This person often is in a "research and development department" and is responsible for the formulation of the basic concepts of new products and/or processes. The development engineer should have a strong background in engineering fundamentals, be able to quickly adjust to new technical situations, and be creative. It also is a great asset to be able to generate enthusiasm and cooperation in others. The development engineer may work longer hours than other engineers because of high interest in current projects.

The *design engineer* is an important coworker of the development engineer. Design engineers do much of the actual design of machinery and processing equipment suggested by development engineers and production people. They use daily the tremendous technology generated over the years in this particular field of engineering. Such

4.4. Designers at work. (Photo courtesy of Ford Motor Company)

engineers often supervise draftspersons, machinists, mechanics, technicians, and others engaged in the final design and construction of the system they originated. A design engineer may be the only engineer a company has and may have to give engineering assistance in many ways.

Often development and design engineers follow their creations into production and actually supervise production operations. However, after the process is running smoothly, they may prefer to return to development or engineering design.

Companies may call in *engineering consultants* to help in developing unfamiliar areas or to improve existing operations. A young engineer may be employed by such a consulting firm. Consulting engineers are frequently on short-term assignments—perhaps lasting only a day or week. Therefore, they charge high fees plus all travel and living expenses. Supervising or self-employed consulting engineers are registered professional engineers almost without exception. They have been examined and licensed by the states in which they work. Consulting engineers may work with quality assurance, production, or design engineers or with top management.

Companies that produce complicated products such as computers, electronic equipment, and special materials and devices must have engineers interfacing with their customers. For that reason, many engineers are salespeople. They have a distinct advantage over nonengineer salespeople in selling technical equipment and services. Engineering training in this case is a tool for selling, much as a knowledge of Chinese would be valuable to an American salesperson in Peking. Salespeople frequently must educate buyers in engineering matters to the point where the buyer can make an intelligent purchase decision. The *sales engineer* especially is in this position when dealing with nontechnically oriented buyers. Sales engineers must be good teachers. Sales engineering requires people who are people oriented rather than thing oriented. They should like frequent travel. Sales engineers usually have above-average engineering income. Inflation has minimum effect on their earnings because commissions rise with prices.

All of the preceding activities require knowledgeable and effective management. Engineers have an advantage in technically oriented industries and usually hold the top management positions.

Of course, many technically oriented companies are founded by engineers.

Those who pursue top management positions in industry must give the company top priority. Time must be carefully scheduled so that family, church, and recreational activities are not neglected. Proper eating and exercise and the avoidance of excess work loads can help to ensure good health. Maintaining an optimistic outlook on life is essential. The potential manager should not be easily upset by what people say and should not take every criticism as a personal affront. It is almost impossible to make important management decisions with an unsettled mind. Not every pursuer of the top rungs of management will reach that goal. Fortunately, there are many excellent jobs in industry for those who do not become the president of the company.

Administrators of large companies are encouraged to become involved in a variety of activities. Often company management personnel work full-time on governmental or industrywide committees and projects, with their salaries paid by the company.

Engineering management encompasses many areas other than the direct operation of the company production lines. An engineer in management needs a strong financial and managerial background, which must be obtained over the years. A fairly large number of engineers will obtain an M.B.A. (master of business administration degree), either by studying at night while working their regular jobs or by full-time study right after they obtain their engineering degrees.

By now you should see that engineers have a wide range of activities in industry. Many civil engineers are generally not engaged in the manufacture of products on an assembly line, but they are engaged in activities similar to those of other engineers. They design, assure quality, supervise construction, and manage large operations that require a range of engineering skills similar to those suggested in this chapter.

The engineering world is broad and complicated. Most engineers are excited by the conversion of knowledge and raw materials to useful services and products, but specialization has sent them down a number of different roads. The next section is a road map to guide your discovery of the fields of engineering.

PART TWO

THE FIELDS OF ENGINEERING

In this section, we discuss the major fields of engineering and relate other fields of engineering to these major fields. Where possible, interviews with working engineers are included. A number of these engineers work with me at Lenox. You should read all of the interviews, not just to learn more about an engineering field but to see what engineers think is important for a successful career.

Although engineering technology is not the subject of this book, keeping the definition of engineering technology in mind may help those of you inclined to that career field. The definition used by the Technology Accreditation Committee (TAC) of the Accreditation Board for Engineering and Technology (ABET) is:

Engineering Technology is that part of the technological field which requires the application of scientific and engineering knowledge and methods combined with technical skills in support of engineering activities; it lies in the occupational spectrum between the craftsman and the engineer at the end of the spectrum closest to the engineer. The term "engineering technician" is applied to the graduates of associate degree programs. Graduates of baccalaureate programs are called "engineering technologists."

5 CHEMISTRY AND MATERIALS ENGINEERING

A number of engineering fields rely heavily on the chemistry and physics of materials and their influence on manufacturing processes and products. The best known and largest of these is chemical engineering. There are over one hundred universities in the United States and Canada that award degrees in chemical engineering. There are over 100,000 working chemical engineers. If you are interested in chemical engineering as a career, there are probably a number of chemical engineers in your locality who would be glad to talk to you about it.

Biomedical engineering, ceramic engineering, petroleum engineering, metallurgical engineering, nuclear engineering, polymer engineering, and mining and geological engineering are smaller fields of engineering that have a chemical or chemical-processing orientation. However, these fields have close affiliation with other fields of engineering too. This is because all engineers work with materials and processes.

CHEMICAL ENGINEERING

Every year about eight thousand students in the United States graduate in chemical engineering. If you are good at mathematics and would like to apply chemistry for world betterment, you should investigate chemical engineering. Chemical engineers apply the laws of physics and chemistry in a wide range of engineering activities. For this reason, a chemical engineer needs a fairly broad education. Chemical engineering students spend about one-third of their time studying approximately equal amounts of chemistry and mathematics. Another third is spent in chemical engineering courses. Added to this are courses in physics and in other engineering fields. About one-fifth of the curriculum is allocated to electives and social-humanities studies.

Chemical engineers work in industries that produce organic solvents, food products, magnesium, uranium, petroleum products, plastics, detergents, aluminum and aluminum oxide, fertilizer, radio isotopes, antibiotics, and many other products. The chemical engineer typically takes the product "as far as the hopper car." After that, other engineers manufacture specific items such as toys, machines, shoes, cloth, ceramics, glass, and millions of other products. (However, chemical engineers work in these areas too.)

If you want to be a chemical engineer, think big! Chemical engineers like to produce material by the trainload—not the thimbleful. When you deal with such large quantities, a savings in production cost of even one-tenth of a cent per pound can have staggering consequences. If a plant produced one million tons of a product each year, the savings in production cost would be two million dollars annually (figure it out). The chemical industries are very competitive, and the chemical engineer must be familiar with economic and financial considerations as well as chemical engineering technology.

To save a few cents per ton of material by improving chemical reactions, modifying equipment such as chemical reactors and heat exchangers, and optimizing maintenance cost, the chemical engineer puts heavy use on control systems and computers. This, with a special ability to design chemical plants, is the chemical engineer's bread and butter. This person needs a special knowledge of construction materials and their proper utilization.

Safety and pollution control are very important factors in chemical engineering. Workers and the public must be protected from foul-smelling gases, explosions, poisoning, and chemical dust. Many chemical engineers perform pollution control and safety functions throughout many parts of industry.

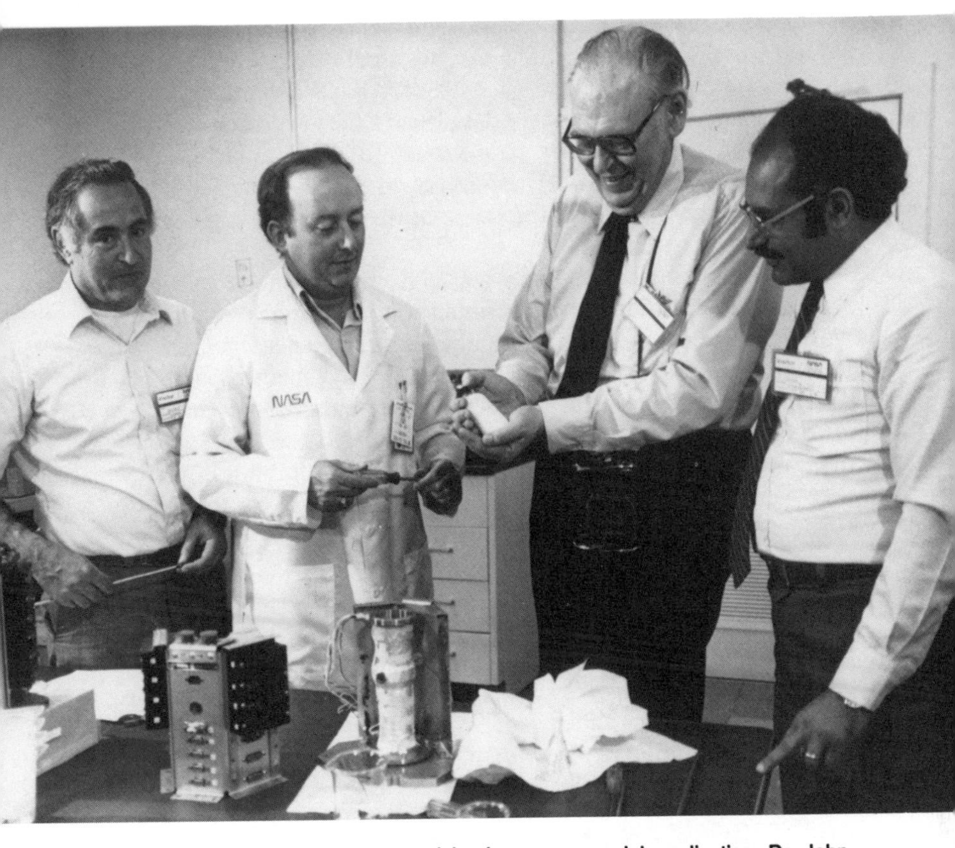

5.1. Latex particles have commercial application. Dr. John Vanderhoff, principal investigator for the monodisperse latex reactor experiment, cradles the first batch of latex particles produced during the STS-3 space shuttle flight in March of 1982. The experiment, conducted jointly by NASA's Marshall Space Flight Center and Lehigh University in Bethlehem, Pennsylvania, has produced large quantities of identically sized latex particles from five to thirty microns during a series of space flights. The beads have possible commercial applications such as calibration of medical and scientific instruments. With Dr. Vanderhoff are other members of the team (l.r., Dr. Fortunato J. Micale of Lehigh, Dr. Dale M. Karnfield of MSFC, and Dr. Mohamed S. El-Aaser of Lehigh). (Photo courtesy of NASA)

Chemical engineering techniques are frequently used in *biomedical engineering*. (Biomedical engineering applies engineering principles to medicine.) The human body is a chemical processing system and the chemical changes that take place in the various "processing units" (organs) can be analyzed according to chemical engineering procedures. By working with physicians, biomedical engineers have developed life-saving monitoring systems that are preserving the lives of critically ill patients.

The versatile chemical engineer can work in energy, pollution abatement, education, agricultural and food products, patent law, plastics and synthetic fibers, surface treatments, instruments, inorganics, transportation (land, sea, air, or space), packaging, pharmaceuticals, medical applications, meteorology (surprise), government, or oceanography. I have added this list to show the many options in chemical engineering. For more information, contact the American Institute of Chemical Engineers or the Junior Engineering Technical Society (see Appendix D).

Sociologists say that persons employed as chemical engineers are among the happiest people who have to work for a living. If you appreciate the conversion of raw materials into useful chemicals by heat, pressure, and clever manipulation, why not join the "happy crowd"? Chemical engineers are employed in transportation, space, manufacturing, government, and many other areas, so you should have no trouble finding one to talk to.

AN INTERVIEW between the Author and Ken Richards, Ph.D., Chemical Engineer

Jones: Ken, you and I met while working together on the synthesis of silicon carbide in the old Ore Dressing Laboratory, School of Mines, University of Utah, in 1956. You graduated that year as a top university scholar in chemical engineering. If you can look back before that time, what first got you interested in engineering and chemical engineering in particular?

Richards: I picked engineering because I had enjoyed the sciences in high school, especially chemistry. My guidance training in high school showed that chemical engineers often became managers in high-paying positions. I wanted to build things and was interested in systems, including economics.

Jones: As I remember, you first worked for an engineering consulting firm. You had many job offers that year. Why did you choose that particular opportunity?

Richards: I wanted to choose a small organization where I could make an early impact. There were a large number of large corporations casting their nets. They told me what they were going to do for me and that later I would be able to contribute to their needs. Small corporations gave me a chance to jump right into decision making.

5.2. This "smog chamber" at Phillips Petroleum Company's research center aids in the solution of many problems of air pollution research. An aim of the Phillips research is to discover principles and techniques to reduce or avoid air pollution. In research on motor fuel, scientists have demonstrated that certain hydrocarbons, such as natural gas liquids components, are relatively innocuous when emitted to the atmosphere while others become smog-forming pollutants at variable rates depending on their chemical constitution. This "hydrocarbon reactivity" concept is a fundamental contribution to understanding the role of hydrocarbons in formation of smog. (Photo courtesy of Phillips Petroleum Company)

Jones: What did you do on that job?

Richards: The company's name was Fractionation Research. I was one of three test engineers. We had large-scale distillation units that would allow us to evaluate any distillation system for large companies. The data was used to design better plants. We had other large equipment for similar evaluations. Data collection, interpretation of data, and reporting were important. This gave me a good introduction into commercial plant operation. It also taught me process control. I learned to write, after gentle tutoring by my boss. Writing is one of the most important tools for engineering.

Jones: Later you took a Ph.D. in metallurgy and went to work at the research laboratory of a large mining company. What led you in that direction?

Richards: I had joined the air force and was allowed to return to graduate school. I had an interest in materials and after graduation worked at the Wright Air Development Center. I felt that the government was not responsible in controlling finances. This, along with a desire to return to process work rather than materials research, led me to Kennecott.

Jones: I assume that you handled major projects for Kennecott. What were the most important and most interesting?

Richards: I think (there were so many things) the one I was proudest of was applying digital computer control to Kennecott operations. The new factory we built could not have operated without it. I had to fight the corporation to do this. Pride came when the system worked and made a new plant possible. R & D [research and development] earned the respect of manufacturing, which allowed more effective R & D and manufacturing.

Jones: You left Kennecott about four years ago to join Kerr-McGee in Oklahoma City. What are your responsibilities there?

Richards: I left Kennecott as vice president of process technology. I had been director of R & D, director of the Technical Center before that. I came here as corporate vice president of technology. That covers everything from exploratory research to process engineering—we design processes, process control systems, do technical and financial modeling, and look at possible new business acquisitions. We are involved in business-driven technical issues. We do not do construction engineering.

Jones: What is most interesting to you about chemical engineering?

Richards: I guess if you summarized what I have said, you would have a good answer. I like to manage large projects that influence the corporation. I like working with the technically wise work forces in the process industries. I enjoy determining the financial impact of technology on the company's business.

Jones: If you had it to do over again, would you still be a chemical engineer?

Richards: Oh yes, no doubt about it! I could have done a lot of things, but the chemical engineering background was excellent. Total systems are what

I like.

Jones: What advice do you have for young people going into engineering?

Richards: First, don't just pass tests, but understand what the philosophy is of what you are being taught. Don't judge your understanding by high grades. Take every opportunity to see how your engineering field is applied. Don't wait until graduation. Try to obtain summer employment in your vocational area. Be general rather than specific. Make the best of economics courses and other courses to see what effect they will have on you being an engineer. It's hard work, so adjust to it psychologically. Consider graduate work early in your education and prepare for that possibility. Try to work in laboratories on campus to make studies more relevant. Don't feel that university work is just a series of hurdles placed there to block your process because it is really the required path to a good-paying, exciting career.

CERAMIC ENGINEERING

The multibillion-dollar ceramic industry converts chemically processed materials (aluminum oxide, etc.) and raw materials taken directly from the earth (clay, sand, etc.) into useful products such as spark plugs, glass, electronic materials, construction materials, tableware, porcelain enamels, refractories to contain hot metals, nuclear materials, abrasives, rocket components, and a multitude of other products. High-temperature processing is the key to ceramic engineering. The products are always inorganic, nonmetallic solids.

Some say that ceramic engineering is high-temperature chemical engineering. This is not entirely true. Ceramic engineering starts where chemical engineering ends. It's true that the chemical and ceramic engineers may be engaged in similar types of activities at certain times, but the chemical engineer's love is the process while the ceramic engineer loves the product.

Ceramic engineers like to relate the properties of solids to their crystal structure and chemical composition. By high-temperature manipulation, they can make important minerals that are rare or nonexistent in nature. These synthetic materials can have a great range of electrical, magnetic, optical, chemical, and physical properties.

From a single composition, ceramics can be made in numerous forms. Carbon can take the structure of diamond for abrasive use,

5.3. Ceramic engineers check out advanced ceramics. *Ceramics for Advanced Applications* is the newest classification of ceramics. (Photo courtesy of Norton Company)

graphite for lubrication, glass (amorphous) for crucibles, fiber for cloth, and special forms for nuclear and other uses.

Glass can be made translucent or opaque, electrical conducting or resistive, and can be changed to forms such as crystalline and ceramic. It can be high-melting for research use or low-melting for electronic applications.

Ceramic engineering attracts students who have an interest in chemistry and physics. Frequently, these students are looking for something a little different from stereotypic engineering. Ceramic engineers often consult with other engineers—especially chemical, electrical, industrial, and mechanical engineers—and must have a broad educational background.

A large number of ceramic engineers work in research and development. They are specially trained to use sophisticated electronic analytical instruments such as the scanning electron microscope and X-ray diffraction equipment.

The glass, refractory, electronic, and technical ceramic companies hire a high percentage of all ceramic engineers. Ceramic engineers do the same things that other engineers do. They sell, supervise production, work in quality assurance, and do consulting.

The National Institute of Ceramic Engineers (see Appendix D) in affiliation with the Accreditation Board for Engineering and Technology (ABET) has accredited the following college and university ceramic engineering programs:

Clemson University
Department of Ceramic Engineering
Clemson, SC 29631

University of Florida
Department of Metallurgical and Materials Engineering
Ceramic Engineering Division
Gainsville, FL 32601

Georgia Institute of Technology
School of Materials Engineering
Atlanta, GA 30332

University of Illinois
Department of Ceramic Engineering
Urbana, IL 61801

Iowa State University
Department of Materials Science and Engineering
Ames, IA 50011

University of Missouri–Rolla
Department of Ceramic Engineering
Rolla, MO 65401

New York State College of Ceramics at Alfred University
Alfred, NY 14802

Ohio State University
Department of Ceramic Engineering
Columbus, OH 43210

Pennsylvania State University
Department of Ceramic Science and Engineering
University Park, PA 16802

Rutgers University
Department of Ceramics
New Brunswick, NJ 08903

University of Washington
Department of Materials Science and Engineering
Seattle, WA 98105

Other universities such as Southern Methodist, University of Utah, MIT, UCLA, University of California-Berkeley and -Davis, Northwestern, and Stanford have undergraduate and graduate programs dealing with the science of ceramics (not to be confused with art ceramics).

After finishing college, ceramic engineers can join the National Institute of Ceramic Engineers (NICE) and gain from its efforts to advance the profession. NICE promotes the professional development of ceramic engineering as an engineering discipline, improves community awareness and understanding of ceramic engineering, and assists in examinations for the licensing of ceramic engineers. Contact NICE or the Junior Engineering Technical Society (see Appendix D) for more information. Also see Materials Science and Engineering in Appendix A.

The "hot thing" in ceramic engineering—the ceramic superconductors—is actually kept at liquid nitrogen temperature. Ceramic superconductors will change your world. You can make it happen!

AN INTERVIEW between the Author and Mike Cooper, Sales Engineer, Leeds and Northrup, North Wales, Pennsylvania

Jones: Mike, you are a ceramic sales engineer. When did you first become aware of engineering as a profession?

Cooper: In high school, but I had been inclined toward engineering—like tearing the family TV apart!

Jones: How did you learn about ceramic engineering?

Cooper: Audiotapes at Pennsylvania State University Library, University Park, Pennsylvania. They were available for all fields of engineering and described some avenues in ceramic engineering [fields], etc. They never mentioned controls [laughter—Mike sells controls].

Jones: When did you graduate?

Cooper: 1980. I started as marketing engineer for Emerson Electric, Pittsburgh, Pennsylvania, after a nuclear engineering position fell through [aftermath of Three Mile Island]. I had done research at Penn State on nuclear waste disposal [firing waste into ceramic structures].

Jones: Then what happened?

Cooper: A "head hunter" [engineering employment agent] saw my resume, which indicated a future interest in marketing. He set up the interview with Emerson. Emerson felt that all sales people needed to be engineers.

Jones: Where did you work?

Cooper: First in Pittsburgh and then Los Angeles. After three years, I was transferred back to Pittsburgh. Things did not materialize in Pittsburgh so I went to Leeds and Northrup.

Jones: In your three years at Leeds and Northrup, what have you been up to?

Cooper: I have been trying to establish a better working knowledge of control and control systems. My ceramic engineering background helps my customers to satisfy their control needs, but I must know the instruments I sell and stay up-to-date in both application and instruments.

Jones: What other types of engineers are selling control systems for Leeds and Northrup?

Cooper: Electrical engineers, mechanical engineers, chemical engineers, and general engineers. There are about a hundred salespeople nationwide, and only two are ceramic engineers.

Jones: Do these engineers interact with each other?

Cooper: Yes, in the district office [each office has thirty or forty engineers] and also with major accounts that cross district lines.

Jones: What are the advantages of sales engineering?

Cooper: You control your own destiny and never get bored; your financial situation is more in your hands.

Jones: Where can a sales engineer position lead?

Cooper: Up through the sales ranks through district and regional management. Some transfer to manufacturing.

Jones: Marketing is a good way to work into the top executive ranks of many corporations.

Cooper: It seems that way.

Jones: What do you see in your future—what would you like to become?

Cooper: At this point, I would like to become a regional manager. Anyone who wants to work into top management through sales must be willing to relocate.

Jones: Do you have any advice for high school students?

Cooper: Students should evaluate the worth of any particular degree they pursue. The choice should not be made on interest alone. The choice should be made on market trends or growth potential.

Jones: If you had to do it over again, would you be an engineer?

Cooper: Definitely. My ceramic education gave me a lot of practical knowledge as well as book knowledge.

AN INTERVIEW between the Author and Ralph Ruark, Ceramic Engineer

Jones: Ralph, you were raised here in New Jersey, right?

Ruark: No. I was born in Maryland. When I was growing up we lived in Maryland; Atlanta; and Allentown, Pennsylvania. We finally moved to New Jersey when I was about sixteen.

Jones: Why did you move around—because of your father's vocation?

Ruark: My father was a sales engineer for a paper company and had to move or was transferred every five years or so.

Jones: Was he a graduate engineer?

Ruark: No, he was a self-taught engineer, a high school graduate.

Jones: Did your father influence you toward engineering?

Ruark: Only indirectly because he wanted me to go to college. He taught me a lot of mechanical and electrical skills and set my path that way rather than directly influencing which way I should go.

Jones: When did you decide to go into engineering?

Ruark: That's difficult. I applied to several schools at the end of high school, and I decided on engineering, I guess, midway through my senior year in high school.

Jones: How did you find out about ceramic engineering?

Ruark: Originally, I enrolled at Rutgers as a chemical engineer, but after about six months of organic chemistry, I decided that I didn't have the aptitude or the interest in traditional chemistry. I liked the ceramics department at Rutgers, and Malcolm McLaren talked me into going into ceramics.

Jones: Dr. Malcolm McLaren is the head of the Ceramic Engineering Department at Rutgers. How many students were in your class when you graduated, and what year did you graduate?

Ruark: I graduated in 1972, and in the ceramics engineering class there were about twenty-two of us, a fairly small class.

Jones: I think that the classes are much larger now, as many as eighty. You obtained a master's degree while you were at Rutgers, right?

Ruark: No, I got into a five-year program, and I decided to get a degree in business. I was going to take my M.B.A. right after I graduated, but instead I just stayed with my bachelor's degree [in business]. I wanted something to supplement my engineering background because at the time there were a lot of aerospace layoffs, and I figured that if I had a little bit of extra learning in business, it would probably help me retain a job in bad times.

Jones: I think that most every engineer that we have interviewed here has said that the business aspects of education are probably the most important thing to go along with the engineering aspects. You graduated in 1972, and when you graduated, what did you do?

Ruark: I got a job with Freeport Refractories Company a couple of weeks after I graduated.

Jones: Are they down in Texas?

Ruark: No, Pittsburgh. I had a great four or five years with that company. I enjoyed it. It was a small enough company that you got to do a little bit of everything, and I discovered more or less by accident that I had an aptitude for combustion technology and kiln operations and process control, and I didn't really want to leave the company. It was more a geographical problem that made us decide to leave.

Jones: Where did you go from Freeport?

Ruark: I got a job with Bickley Furnaces as their laboratory manager in combustion technology and worked for Bickley for about nine years out of Philadelphia.

Jones: Besides running the laboratory, you also were the kiln commissioning engineer.

Ruark: I started as lab manager and after a couple of years was combustion engineer, and then I was made manager of the commissioning department. That was a great job, a very interesting job because of the travel. I had a group of six or seven engineers, and we would start up kilns and/or whole factories all over the world.

Jones: That kind of kiln engineering takes a lot of long hours, a lot of bad food, right?

Ruark: It was a great job for a lot of reasons. First, you learn more in a year (for me, anyway) than the previous ten or fifteen years because you learn under duress. When you are alone, in a foreign country especially, and have to solve a problem, there is nobody there to hold your hand—you just do it. You get accustomed to working the long hours, and soon it becomes a very exciting job because you take a kiln in which perhaps the electrics don't work and maybe the hydraulics don't work and the combustion system is not set up, and you make it a functioning unit over a period of two weeks to maybe six months.

Jones: How did your education prepare you for that type of activity?

Ruark: I think—like most undergraduate education—I learned methods of thinking and accomplishing things more than anything that I learned specifically in subject content. I learned how to approach problems and how to solve problems. I think, because of that, the method rather than the actual raw data was what I got out of Rutgers.

Jones: Did you have adequate education in kilns and combustion systems to prepare you for your work, or did you have to learn after you got out of school?

Ruark: There was no combustion training at Rutgers. You learned in unit operations a smattering of different processes, and that was helpful because one week I might find myself in a sanitaryware factory and then maybe technical ceramics capacitors, insulators, or abrasives, or sometimes heat-testing metals, and so if you had a little bit of knowledge about some of the processes it helped going in. You learned a lot more on the job in a real short time.

Jones: I think that when I was a student at the University of Utah, I received a lot less training in kilns and combustion than what I taught as a professor at Iowa State University. We had a pretty intensive program there in kiln and factory design. But people don't do that as much anymore. They are more fundamental. We tried to give the students both the chemical engineering unit operations and also a practical application of designing a kiln and actually doing combustion studies in the laboratory. I know that Alfred University, for example, has combustion training.

Ruark: I think it is crucial because thermal processing is common to just about every ceramic process. I graduated with really very little idea of how combustion systems work or electrical elements or SCR's [silicon-controlled rectifiers] — it was a big hole in my curriculum.

Jones: I think it is true of all engineering fields that you will not get enough information in college for the specialties you get involved in when you get out. But you learn enough of the fundamentals so that you can train yourself or easily be trained by others to perform adequately when you are working in the field. You are involved in process and product or have been. The final product is really the thing that ceramic engineers are interested in, and [as I pointed out elsewhere in the book] chemical engineers love the process, but you have got to remember too that you don't get a product without the process.

Ruark: It is hard to love oil or plastic, but it is nice to be involved in some ceramics.

Jones: Even some of the most common ceramics are quite beautiful works of art if you don't think of the utility of them. Now that you are working here at Lenox as the director of kiln technology, you have worked under a lot of stress caused by a heavy project load and also a lot of travel. How have you been able to cope with this?

Ruark: I think you never know what you can do until you have to do it. I think that stress is mainly internally, not externally, generated, and so, even though my project assignment load is heavy, generally I generate what I feel about it — the tension I feel — myself because I put my own requirements on each job that I do, requirements that are usually tougher than those assigned to me. So, to cope, sometimes I find it much easier to work longer hours than to be concerned with things that I should be doing, and I try to balance that by keeping the weekends exclusively for my family unless I am traveling. I probably spend more time with my wife and children than most people do on the weekends and in the evenings.

Jones: This is important, I think, for young engineers to realize — that during their career they will be under stress because of the difficulty of the work. It begins in college with a heavy course load in some really tough subjects, and, as Ken Richards said in his interview [in the chemical engineering section of this chapter], you have got to learn while you are young to adjust psychologically to the burden that is in front of you. That is a problem that every engineer will encounter, especially those that are managers. Just about every engineer will do some type of management work before his or her career is over. It will start out early, maybe with a couple of technicians to supervise. At the end of a career, a manager might be running a corporation or several corporations. Everything that you learn along the way that helps you to perform better is going to help relieve stress. One way to relieve stress is to push work off on other people.

Ruark: In my department, the only person I push on is myself at the moment. What helps to relieve stress for me is completion of a project. Most of the projects that I work on are longer-term, so you don't get that instant gratification that maybe you got in college when you worked hard on a project and you got a grade in a month or two. In some of the longer-range projects, you work hard for so many months that it is easy to lose sight of the fact that, yes, it will be done one day. One way to counteract that — something that I do even though it places an additional work load on myself — is I try to stay involved in shorter-term plant projects. The more projects you have going on, the more interesting it is, and the more likely it is that one of them is going to be completed and you will get to measure how you are doing, and I like it like that. It keeps you busy as hell!

Jones: That is an important point, and it again reflects back to what Ken Richards said — that even as a student you probably should try to keep a broad view of your activities, and that way you can see the light at the end of the tunnel. The point that I thought was interesting as you reflected on grades was that Ken, a top scholar at the University of Utah and the top engineering student, said that students mustn't judge their performance by the grades they are receiving in class, but they should judge it by their understanding the philosophy of the courses and how they will fit into their lives later when they are operating engineers. He, of course, expressed the same thing you did, as did others, that business aspects — if you know something about accounting and business — make your life in engineering a lot easier.

Ruark: One reason that you work as an engineer (among others) is to solve problems, and as you solve problems, the solutions have an impact on the bottom line of the corporation. The better that you understand what the makeup of that bottom line is, the better you can approach your job. Sometimes it is easy to get lost in a sea of numbers and a sea of projects. If you do something that has real value and you understand how it fits in with the corporate P&L (profit and loss), it gives you a little extra incentive to do a good job. Not that you really need incentive, but it is always nice to know if you are working hard, putting in a lot of hours, that it will have an impact somewhere along the line, and it is good to know what that impact will be in dollars.

Jones: I think that is true. I think Ken Richards was the only engineer that mentioned in interviewing the pride that he develops by accomplishing difficult tasks. The other day we had a chance to fly to Europe on the Concord, and both of us were very interested in the statistics of that aircraft and the performance. But have you thought much about the engineers who had to design that? They were certainly on the edge of technology, and they were not only working for the companies that were trying to build it, but governments were also involved. Can you imagine all of the stress and bureaucracy and all of the things that they had to go through to accomplish the design of that aircraft? I don't see much difference between the Concord and the space shuttle after they have taken off the ground, except that the Concord has one capability that the space shuttle doesn't have. That is, it can make a slow takeoff—if you call 250 mph slow! After we took off in the Concord, the retrorockets went blasting off and we were up to Mach 2 and then back into the heavier atmosphere—of course, the temperatures generated by friction aren't quite as great coming in from 58,000 feet (or higher) as they are from space—we could feel the heat in the hull, and we had a very fast controlled "crash" for a landing that wasn't much different from a landing from space. If you get the picture, those engineers were on the threshold of science and engineering, and in every field of engineering—if you're doing anything really new—you are going to be on some threshold. I think that we are probably on some thresholds right now with our company.

Ruark: I was just thinking, when you talk about pride—I think anybody that does good work has a level of pride in their work. As you know, we have just been working on justifying a project so I am a little more focused in on the bottom line aspects, but in the next week or two we will be starting a big kiln that I had some involvement with, and the real payoff is seeing some equipment that you participated in the concept of and the design of operate, seeing it do a good job. That's fun. I have one of those things coming up—really two of them coming up—in the next two weeks, and, although it will be long hours in commissioning and a lot of headaches debugging and so on, if it works as well as we hope, it will be a lot of fun too.

Jones: Well, that's for sure! When I go down to our factory in North Carolina that we built a few years ago, it is really interesting to me to see the development of the people and their ability to put out really top-notch products with very little experience compared with the workers in our other factories.

Ruark: The Oxford plant is funny, and that is where I think engineering foresight is needed to help some of the accountants and financial people. In our case, I think they felt they were going to build a plant and immediately were going to see a payback. It actually took two years to build up the work force skills and to modify and debug the equipment to get the payoff of higher yields and higher quality. That is where an engineer can guide the financial manager on any project—to advise him that the buildup isn't going to be immediate, it is going to take time, it is going to have to phase in. I think the financial person often doesn't know that. An engineer can play a very important role in setting the expectations for a project up front rather than at the end where everybody thinks it will do *X* and it actually doesn't do *X* for a couple years.

Jones: That's right. As you know, I went to an international technology conference in Miami a few weeks ago. After I got there, I found that what this conference was really all about was to try to merge business education with technical education, and the only way they were able to do that was to take engineers who had experience and have them teach in the business college. Only that way could they get the technological aspects explained properly in the classroom by people who had been trained in that area and yet could relate to the business aspects.

Ruark: That makes a lot of sense. One of the things that a successful engineer has to be able to do is communicate effectively. Because, although you may think of yourself as a specialized engineer, in fact, you have to deal with all levels of your own company, all levels of possible clients or vendors, and you have to be able to express not only the process but the impact of the process on people who work within it, the impact of the process on the bottom line for the businessperson. You have to be able to deal with white collar people and blue collar people, and, although you think of yourself as an engineer, you really have to be a go-between for a lot of segments of a lot of different areas within and without your company.

Jones: Again, we can go back to what Ken Richards said — that the most important thing he learned first after graduating from school was how to write. He said that at the end of his first project he wrote what he thought was the most beautiful technical description of a process including the collection of data and the interpretation and the conclusions. He thought that he was going to be praised and rewarded for having written that document. He said that when it came back, it was one solid mass of red corrections from his boss, who very gently took him through each problem and taught him to write over a period of time. I remember my boss doing the same thing with me — helping me to write — and you may have seen me doing it with young engineers at times, or sometimes with older ones. To be able to write is really important because if you can't explain to someone else what you have done, you are going to have difficulty in ever having someone else accept programs that you have developed.

Ruark: Absolutely, and that goes hand in hand with the ability to make good oral presentations. Usually the men and women who decide whether a project goes forward aren't necessarily going to be engineers. They are going to be financial people or the presidents of companies; they may or may not be engineers, and you have to explain things without being condescending. You have to be able to explain what it is all about in some technical terms, but generally you have to be able to tell them what the story is about without a lot of technical gobbledygook, and sometimes that's tough.

Jones: That was a good point made by Mike Asche, the engineer we interviewed about agricultural engineering [see Chapter 9]. Early in his career he was promoted because of his ability to communicate with John Deere dealers, John Deere salespeople, John Deere mechanical technicians, and farmers and other customers of John Deere. That — his ability to communicate — plus his technical ability has earned him the job of a territorial manager for John Deere. Last night I listened to a book review by Steve Allen on public speaking, and, although the principles of public speaking are really quite simple, you need to know what you're talking about. You shouldn't try to speak in any form that is not natural to you. If you're not used to telling jokes, you shouldn't tell jokes. If you are used to telling stories, tell stories. You can do the things that are natural for you to do, and simple preparations like this are all that is required to learn to speak. However, few people do public speaking when they are young, and, if they don't have the opportunity either in church or in college, they become very frightened when they stand in front of an audience. But these fears are easily overcome. If you prepare properly, know your subject, don't put on any airs, and just talk naturally as if there were only one person in the audience who is your friend, then you will be able to communicate. That is something that people have to learn to do in engineering or any other field that is a part of business.

Ruark: I took a course in public speaking perhaps five or six years ago because I had a big seminar to do, and I was very nervous. It was certainly among the most valuable courses I have ever taken. I am always nervous before I speak to a group, and it is size independent. If there are more than twenty or thirty people, then I am nervous, but if I prepare, after about the first sentence I feel really good about it. If I don't prepare, I am nervous all of the way through.

Jones: That's what Steve Allen says. He says there is something about your voice that relaxes you when you hear yourself talking and you see that everybody doesn't get up and walk out of the room. You need to know your subject. You are known as one of the best oral communicators in the company, and, of course, our division president is top-notch in public speaking skills.

Ruark: He is good one-to-one, too.

Jones: Yes, he is good one-to-one, and these skills are important. We can't overstress communication skills. What kind of advice would you give to high school and university students as they pursue an engineering career?

Ruark: I found engineering to be completely different from what I expected upon graduation from college. Based on my experience, engineers have to be very flexible. Even though you major in a particular engineering field, don't be surprised if you end up working in sales or marketing or project development in some other field of engineering or maybe in science. In the job

I have, there is some engineering and there are some things that don't have anything at all to do with engineering. Don't plan your career to the point that any surprises will upset you. You sort of have to roll with possible changes and opportunities that come along. Engineering, at least for me, isn't only sitting at a desk, grinding out solutions to problems on paper or on a computer. It is much, much broader than that. My job can go from one day giving an oral presentation to officers of the company and then the next day having on coveralls and holding a pair of pliers in my hand as I adjust burners in a kiln. It runs such a wide range, and it is not what I expected. But oddly enough, the variety is very enjoyable.

Jones: You interrelate with a lot of other types of people in your job, many of them engineers. Did your training prepare you for this type of activity?

Ruark: Unfortunately, at Rutgers, the College of Engineering is situated on a separate campus, and you tend to relate only to other engineers or engineer types. Fortunately for my background, I also majored in business, which was more liberal arts oriented. It was a big help because I associated with engineering students half the time and liberal arts students the other half of my time.

Jones: As you know, because of my position in the company, I have to work with every different aspect of the business, and one of the things that I enjoy is marketing—working with product development and marketing areas. Do you have a favorite thing that you like?

Ruark: I think, for me, my favorite aspect of my job is work on the development of a project, get the required quotations, have meetings with the vendors, and then bring it to a contract document, monitor it, and see everything we've talked about in the last months become a reality and actually fire a product to high-quality standards.

Jones: I think that is the essence of engineering, and most of the engineers have said that that is the thing that they like best about it. That they are actually able to see the changes that they have caused.

AN INTERVIEW between the Author and Lisa Merget, Ceramic Research Engineer

Jones: When did you first become aware of engineering?

Merget: I finished high school at age sixteen. There, I had some exposure to engineering guidance. After graduation from a community college (two-year program, associate in science), I decided to go into engineering because of a strong orientation toward science and mathematics.

Jones: You are a ceramic engineer. How did you learn about the field?

Merget: I read about it at a Rutgers University transfer student orientation (along with other information on engineering).

Jones: When did you start your ceramic engineering training, and why did you select that field?

Merget: I started during the sophomore year (second semester) along with all engineering students at Rutgers.

Jones: Then you actually went through a normal four-year program plus your junior college.

Merget: Yes. I transferred about thirty credits from junior college, which allowed me to work in the Center for Ceramic Research starting during my sophomore year.

Jones: Then you were able to gain hands-on experience because of a lighter class load.

Merget: Yes, but the junior year was tough because that is when the hardest ceramic engineering courses are taught.

Jones: It's interesting (to me) that I also started into ceramic engineering because of a campus job. How did you experience being a woman in an engineering program?

Merget: Our class had a large number of women, three of whom were the top students. One of these is working at DuPont, one at NASA, and the other one is in graduate school.

Jones: You graduated during a tough year for new graduates in terms of employment. How did you get your first job?

Merget: You hired me! [laughter]

Jones: We interviewed at Rutgers, right? And that is how we got together. You have had a number of assignments at Lenox. How have you adjusted to your position in research and development?

Merget: It was an easy transition, having worked in the laboratory at Rutgers.

Jones: Have you had any problems because of your being a woman?

Merget: More problems because of my age [twenty-three years old] than being a woman. There was nothing I couldn't handle.

Jones: You are now working with the scanning electron microscope. How does this fit in with your Rutgers training?

Merget: Beautiful. It is like "icing on the cake." It gives me a break from my regular work, and I like to go in there and do it.

Jones: Would you advise other women to go into engineering in light of your research work and also your interaction with manufacturing and management?

Merget: Most definitely, yes—here at Lenox! And women should certainly be engineers rather than factory workers. I worked "B" shift [4 P.M. to 12 P.M.] in a factory one time. It's better to be an engineer.

AN INTERVIEW between the Author and Jill Badiavas, Quality Engineer.

[Badiavas had been a process engineer but lost her job when a factory was closed.]

Jones: Jill, you went to Rutgers, right? When did you graduate?

Badiavas: I graduated in ceramic engineering in 1981.

Jones: How did you become interested in engineering?

Badiavas: My father and many of the other men in my family are engineers — electrical, mechanical, etc.

Jones: What about the women in the family?

Badiavas: No engineers, but my aunt is an astronomer.

Jones: When you started school, which engineering fields appealed to you?

Badiavas: I liked chemical engineering and ceramic engineering.

Jones: Did your education prepare you for your future work?

Badiavas: I learned the basics — how to be methodical, how to solve problems, and where to look for solutions.

Jones: In other words, they taught you to think, which was more important than mere facts.

Badiavas: Yes, the facts you can find.

Jones: Did you enjoy college?

Badiavas: Yes, I had fun and enjoyed all aspects of college. But I had a tough time at home due to the death of my father, and I had to work hard at college.

Jones: What did you do after college?

Badiavas: I worked for Owens Corning Fiberglass as a process engineer.

Jones: What was your first assignment?

Badiavas: I worked on a roofing mat line. I worked closely with R & D, coordinating production trials. I did scheduling, economic analysis, trouble-shooting, etc.

Jones: Did you move on to other departments?

Badiavas: Yes, and the responsibility increased as I was held responsible for operations and failures. It was fast paced.

Jones: Was there too much pressure?

Badiavas: No, I liked it. I was never bored, and you work better when you are busy.

Jones: You were there for over six years.

Badiavas: Yes, I started working at Lenox China last summer as a quality engineer. I like this job because I like to supervise people and interact with them, but I would much rather be in process engineering where I could be closer to the technology and not just the product. I would not like to stay in this position.

Jones: What advice do you have for young people who might want to prepare for an engineering career?

Badlavas: It is the same for men and women. You must work hard and study and compromise between alternatives and possibilities. You can't do everything. You have to pick off the things you can do and do them.

METALLURGICAL ENGINEERING

Metallurgical engineers are engaged in the recover of metals (see "Minerals Engineering" in the geological engineering section in this chapter), production of metals (steel, magnesium, aluminum, copper, etc.), fabrication of metals (aerospace, automotive, appliance, machinery, etc.), manufacturing (chemical, food, paper, etc.), consulting, and research. Metallurgists create new metal alloys to resist erosion, corrosion, high-temperature deformation, impact, and fatigue. Special alloys are developed for areas such as power distribution and nuclear applications.

Since metals are important in all human activities, one can easily see how broad the field of metallurgy is.

The science-oriented metallurgical engineer studies modern research methods and instrumentation and is also deeply interested in the impact of processing on the metal product. Heat treating and control of the constituents in a particular alloy are of special interest. Some prefer to study metallurgy as a science rather than as a field of engineering, and many schools give the student this alternative.

Areas of metallurgy include extractive metallurgy, physical metallurgy, powder metallurgy, nuclear metallurgy, and others, but I believe that I can give you a better feeling about the field from personal experience. Both of my grandfathers were miners.

My paternal grandfather was sent to the lead mines in North Wales as a child. He held the drill bits for the powder men. As they hit the drill bit with the sledge, he turned the bit. Later, he became a powder man and worked in coal and metal mines in the United States. He saw John L. Lewis leave the mines in Pennsylvania to become "educated." Later, in Colorado, my grandfather was "leaded" for the second time and had to leave metal mining. His first case of lead poisoning, as a teenager in Wales, had nearly cost him his life—there was no OSHA or John L. Lewis around in those days.

My maternal grandfather worked in many mining towns in Utah. My mother was born in Silver City (now sagebrush) and moved from one tent city to another before settling in Bingham, Utah, the site of the largest open-pit copper mine in the world at that time.

As a boy, I used ore samples from the mines my grandfather had worked as crystals for my home-brewed crystal radios. I knew nothing about semiconduction in those days but marveled that these special metallic rocks could, with a wire "cat whisker," rectify radio waves.

Many of the men in my neighborhood worked at the Bingham copper mine or the refineries at Magna, Utah. Some worked in the steel mills near Provo, Utah. I watched the economy change by observing how the metallurgical industry was doing. The indicators were layoffs and recalls of workers.

As I grew older, some of my friends began working in metallurgical operations and told me what they were doing. I started hearing about all kinds of exciting processes. I had a chance to visit operations, and after high school I worked as an outside contractor in some of the metal refineries (as well as oil refineries and sugar refineries).

This education was interrupted by the Korean War, and I didn't think about it again until I obtained a part-time job in the Ore Dressing Laboratory, School of Mines and Mineral Industries, at the University of Utah. Here, work had been done to extract uranium metals from ore.

While majoring in ceramic engineering, I met the world-renown extractive metallurgist, Dr. Milton Wadsworth (who became dean of the school). He was (and still is) a great teacher, and because I took five years rather than four to obtain my B.S. degree, I was able to take a number of metallurgy courses. (Later I took the major course work for the Ph.D. degree in metallurgy but declared it as my minor because of my decision to return to the ceramic industry. My Ph.D. is in ceramic engineering with a metallurgy minor.)

The Ore Dressing Laboratory was an exciting place during the three years I worked there. Some days I was running the equipment as part of my job and other days as a student. "Uncle Milty" (we never called him that to his face—we called him Milt) taught us how to extract metals from low-grade ores both in the classroom and in the laboratory. We saw graduate students working on real projects daily. Ceramic engineering students and metallurgy students worked side by side doing about the same things but with different materials.

We learned that there was a lot of surface and physical chemistry used in metallurgy. We learned how metals are fabricated, and we visited actual operations. We saw gold, copper, and steel fabrication — always starting with the ore and then going through the complete refining operation.

We learned how metals change by heat treatment and by physical treatment. We used X-ray diffraction along with other sophisticated equipment to study structures. We learned how to harden metals, how to make them resistant to corrosion, and how to make them malleable.

I got so wrapped up in metallurgy that I took several graduate metallurgy classes as an undergraduate. Five years later when I returned to the university for graduate work, I studied physical metallurgy, surface chemistry, solid state physics, nuclear metallurgy, physical chemistry of metallurgy, and more X-ray diffraction, all in the metallurgy department.

As a ceramic engineer, I have worked with a number of composites that have metals bonded to ceramics. I have specified metals for high-temperature or corrosive processes. Engineers must deal with metals. In my opinion, every engineer should study some metallurgy or materials science as an undergraduate.

The advancement of engineering continues to rely on the simultaneous advancement of materials, be they metals, ceramic, or polymer. You can study materials science or specialize in one of the three material areas. If you want to be a scientist rather than an engineer, consider the following relationships.

Scientist	*Engineer*
Chemist	Chemical engineer
Ceramist (not art type)	Ceramic engineer
Metallurgist	Metallurgical engineer
Polymer scientist	Polymer engineer

The field of metallurgical engineering will participate in the great demand for engineers in the next decade. If this sounds like your cup of tea, contact a metallurgist living in your area. Also contact the American Institute of Mining, Metallurgical, and Petroleum Engineers or the Junior Engineering Technical Society (see Appendix D).

NUCLEAR ENGINEERING

The field of nuclear engineering is concerned primarily with nuclear power generation. Much of the technology of chemical engineering in relation to reactor design is involved. Nuclear reactions take place under severe conditions, and there are numerous problems to be resolved with respect to materials for both fission fuel and construction.

Since this field relies heavily on the skills of mechanical, electrical, ceramic, metallurgical, and other engineering fields, some schools prefer to take graduates from other engineering fields and train them in nuclear engineering at the graduate level. Nuclear engineering programs are often under the direction of other engineering departments such as chemical engineering or mechanical engineering.

Nuclear engineers will have an important function in the decades ahead. Those who like tough problems will find their fair share in nuclear engineering. One of these problems is to help supply consumer and industrial power. The national demand for electrical power is doubling every ten years. Much of this power will have to be supplied by nuclear energy.

If we are to explore space, we will have to rely on nuclear power. The propulsion systems used to explore the moon are too cumbersome for the exploration of the planets. Compact propulsion systems are being designed by nuclear engineers.

Water is an interesting commodity. Modern society uses a tremendous quantity for drinking, sanitation, irrigation, power generation, and numerous industrial operations. Future water needs will have to be supplied in part by desalination of sea water. Nuclear desalting plants should play an important function in water purification—and in the production of chemicals taken from the sea.

Nuclear engineers will play an important role in food production and food processing. They will also be engaged in the fight against pollution and will continue to play a vital role in medicine.

Transportation will also be an important emphasis of nuclear engineering. Nuclear ships can "sail" for up to five years without refueling. Roadways and waterways can be prepared by using nuclear blasts. Mining of low-grade ores may be made possible by the use of nuclear excavating.

You can see that there is a great future for nuclear engineering. The United States Atomic Energy Commission has many educational booklets related to nuclear engineering. The address is USAEC, Division of Technical Information, P. O. Box 62, Oak Ridge, TN 37830. The booklets are free for the asking.

5.4. Oil displacement experiments performed in this specially built one-hundred-foot folded tube are a part of Phillips Petroleum Company research that has led to new methods for recovering a higher percentage of oil from underground rock formations. (Photo courtesy of Phillips Petroleum Company)

PETROLEUM ENGINEERING

Petroleum engineers are essentially specialized chemical engineers. Petroleum engineers try to locate and develop new sources of fuel from coal, oil shale, and other sources. They also try to improve existing fuels and methods of fuel production. Petroleum engineers work in many parts of industry where the development of new fuels or the improvement of the utilization of fuels is important. Needless to say, petroleum engineers are in great demand during energy shortages.

Many petroleum engineers work worldwide for integrated oil companies. Their work can be restricted to exploration and production or to processing and marketing. As with other engineering fields, work is not restricted to industry. Many petroleum engineers work for universities and government.

Petroleum engineering is offered from the University of Alaska to the University of Wyoming (see Appendix A). The industrial demand for petroleum engineers is cyclic, but the future is certainly strong. Exciting positions in research, production, and management will continue to provide challenges to future engineers.

GEOLOGICAL AND MINING ENGINEERING

If you have an interest in geology along with engineering, you should investigate the fields of geological and mining engineering, which deal with the exploration and recovery of natural mineral resources.

Geological engineers are primarily interested in mineral exploration, flood control, earthquake-damage prevention, and building-foundation design. They find employment with large and small corporations that seek new sources of petroleum and other minerals throughout the world. Many work with governmental agencies while others are self-employed as consultants.

Mining engineers must be versatile, with a primary interest in operation, design, and manufacture of material processing and transportation equipment (earth-moving equipment, etc.). They may be involved in the mining and concentrating of low- or high-grade ores. They must know the fundamentals of mine management, including the control of dangerous gases, structures (tunnels, etc.), and exploration techniques, and have a good background in economics.

Minerals engineers (metallurgical engineers or process engineers) direct processes separating valuable minerals from worthless materials. A number of technical areas are involved including crushing, grinding, chemical treatment, separation, electrolytic deposition, and other processes.

Coal mining is an important part of minerals engineering since 82 percent of our energy reserves are in coal. Other fields include energy, metals industry, fertilizer production, cement, quarries, consulting, teaching, and government. Accredited programs in minerals fields leading to degrees in engineering or engineering technology are offered at the educational institutions as listed in Appendix A and Appendix B under Geo-Engineering, Geological Engineering, Geophysical Engineering, Minerals Engineering, Mineral Process or Processing Engineering, and Mining Engineering.

6 BUILDING AND CONSTRUCTION ENGINEERING

Engineers involved in the construction of buildings, dams, stadiums, and roads are better known to the public than are most engineers. They are frequently caricatured in cartoons as wearing hobnailed boots, Levi's, plaid shirts, and hard hats. A transit is nearby—often contrasted with a bulldozer or crane.

Civil engineers and architects form the largest groups in the construction area. Civil engineering is divided into a number of important branches due to the complexity of public and private systems and facilities. Preparation for construction engineering requires a special curriculum frequently administered under civil engineering departments. It will be discussed separately.

Architecture is a profession so closely related to the building and construction engineering fields that it is included here for reference.

ARCHITECTURE

If you would like to provide a pleasant, humane, efficient environment for others, perhaps architecture would please you. Studying architecture offers you a broad educational background in the arts and sciences plus a technical background in engineering skills.

Architects typically obtain a B.A. or other bachelor's degree after four years of study. This degree does not allow them to join the

more than thirty-thousand registered architects in the United States. However, a person with such a degree can find a career in business or construction.

To become a registered architect, a person must obtain a professional degree in architecture. This is usually a baccalaureate (B. Arch.) or master of architecture (M. Arch.) degree. (Most B. Arch. programs are being phased out, however.) Students are encouraged to work with an architect for one year before their last year or two of schooling. An additional two years of schooling are generally required to complete the first professional degree (M. Arch.). Thus, if a person works for one year in the midst of formal education, she or he will spend about seven years in becoming an architect.

Architects design urban communities and industrial facilities in addition to homes, schools, and churches. They try to design structures that encourage the growth of human values. Foreign travel and study are encouraged to give architects a broad view of the past history of architecture and to help them to be innovative in their own environments. You can see that the architect needs the eye of an artist, the technical background of an engineer, and a good liberal arts background.

Architects are true professionals because they almost always work directly with clients whether a client is building a home, a planned community, an office building, or a factory. This important client-architect relationship determines how the architect will apply knowledge of design, space, lighting, heating and cooling, structures, and many other construction factors.

The architect wants to design a pleasing creation but can never forget the functional aspects of the design. A school is different from

6.1. NASA truss structure for space station, Houston. Les St. Leger, manager of the Structural Test Lab at NASA's Johnson Space Center, checks a model of a "tetratruss" structure of the tetrahedron design. A full scale graphite/epoxy truss cell of the tetrahedron design was built by McDonnell Douglas Astronautics Company and is under consideration by NASA for use in making platforms for the space station to be built in the 1990s. (Photo courtesy of NASA)

6.2. Southern Nevada Water Project, Stage II. Berney Tafoya, on the Engineers Rotation Training Program, (on field assignment at the Southern Nevada Construction Office in Henderson, Nevada) is shown checking the gap and the alignment between the two manifold sections to bring water from Lake Mead to the pumping plant. (Photo courtesy of Bureau of Reclamation)

a hospital, and a church is different from a mausoleum. Also, the architect must ascertain that the contractor selected to build the structure is competent and that the contractor builds the structure according to the architect's drawings. After construction, the grounds must be properly graded and landscaped.

Many architects are excellent artists. Some individuals who could become good architects make a self-judgment on their own artistic talent by comparing their ability with that of experienced architects. They then decide not to go into architecture. This is sad because many people can be taught to do acceptable art work in accredited architecture programs. If you have an interest in architecture, discuss the possibilities with working architects and visit a university architecture department in your own state. Only then will you be able to make the correct decision.

This is the age of CAD, CAM, and CAE (computer-aided design, manufacturing, and engineering). So use the computer, hire a draftsperson or artist, or learn to draw!

CIVIL ENGINEERING

Do you want to cut a tunnel through the Rockies? How about constructing a superhighway in Brazil or Canada or perhaps a water-treatment plant in Tucson? If this appeals to you, then you should consider civil engineering as a career. The American Society of Civil

6.3. Southern Nevada Water Product, Stage II. Berney Tafoya, on the Engineers Rotation Training Program, (on field assignment at the Southern Nevada Construction Office in Henderson, Nevada) is shown checking the flange for the vertical position, which will be connected to one of the pumps at the pumping plant. (Photo courtesy of Bureau of Reclamation)

Engineers has over 67,000 members.

The design analysis of bridges, buildings, dams, tunnels, and structures requiring the calculation of forces created by loads, wind pressure, and earthquakes is done by *structural engineers.* Structural engineers design conventional and nuclear power plant structures, buildings, towers, and bridges under a wide range of conditions. They build on mountaintops or in swamps; over, under, or in rivers, lakes, and oceans; and against extreme conditions such as flooding, tornadoes, or severe earthquakes. The structures they design must be compatible with the planning of architects, highway engineers, and the laws governing the geographical areas in which they are building. Structural engineering is a varied and exciting career taking many engineers to all parts of the globe.

Transportation engineers design, construct, operate, maintain, and administer rural highways, urban streets and expressways, railroads, waterways, harbors, airports, pipelines, and other transportation systems. This is the largest of the civil engineering fields. Most transportation engineers build roads and highways. Since highways transverse a wide variety of terrain, highway engineers are faced by a variety of problems generated by such factors as soil conditions, mountains, and rivers. Highway engineers also must consider community planning, proper use of land, and future conditions.

The provision of safe and ample water supplies is the responsibility of the *sanitary engineer.* This person is also engaged in the design of sewage treatment plants and other facilities required to maintain the health of a community. Sanitary engineers contribute to the battle against air pollution. The design of sewage treatment plants is accomplished by the combined efforts of sanitary engineers and biochemical engineers (may be one and the same person). It helps a sanitary engineer to have a basic knowledge of bacteriology and chemistry.

The *hydraulic engineer* (hydrologist) works with the technical aspects of the application and conservation of water supplies. Irrigation and drainage are important aspects of this type of engineering. The research of the hydraulic engineer is very important to other civil engineers. The size of dams, power plants, bridges, irrigation and drainage systems, and other structures depends greatly on the availability of ground and surface water and on the time periods when water is most and least available.

Soils engineering is concerned with the engineering properties of soils and their relationship to the construction of buildings and dams. Soil engineers obtain the information needed to build foundations for buildings and the many other structures of civil engineering. You may have watched soil engineers take soil samples by drilling on future building sites in your community. These samples are tested for load-bearing ability and other variables. The structural engineer or architect designs the foundation according to the results of these tests.

Surveying and *mapping* are two areas important to all branches of civil engineering. You have probably seen civil engineers surveying roads and highways. A wide variety of techniques is used to do surveying, including aerial, electronic, and the familiar optical techniques. It can safely be said that no civil engineering structure of any consequence is planned or constructed without the help of surveyors.

AN INTERVIEW between the Author and Walter Smith, Civil Engineer

Jones: How did you become interested in engineering?

Smith: By watching construction projects—seeing people build things.

Jones: When did this occur?

Smith: As a child.

Jones: How did this lead to engineering?

Smith: I also liked mathematics and drawing.

Jones: Was this in high school?

Smith: Yes, I wanted to understand *why* things were done the way they were, not only *how* things were done.

Jones: Did you start out into civil engineering on first entering college?

Smith: Yes.

Jones: Which school did you go to?

Smith: Rice University.

Jones: And that's in Texas, right?

Smith: Yes, Houston.

Jones: Did you finish in four years?

Smith: Yes.

Jones: How did you finance your schooling?

Smith: By scholarships and working summers. Probably some of my best education came from working in the summers.

Jones: I assume you worked in construction during the summer.

Smith: Yes, but I worked for an engineer.

Jones: Upon graduation, what did you do?

Smith: I worked in land development, including roads, sewer lines, drainage, etc.

Jones: How did your education prepare you for that?

Smith: It gave me the technical background.

Jones: Are you still in that line of work?

Smith: Yes, but I work mostly on my own. I did have my own consulting business, but now I have one exclusive client.

Jones: If you had it to do over again, would you be a civil engineer?

Smith: Yes, the exact same thing.

Jones: Most engineers give similar answers to that question.

Smith: They should if they chose the field for the right reasons. It would be a shame to have to do work you don't enjoy.

Jones: What do you like most about civil engineering?

Smith: At the end of a job, I can see changes that I made come about in the environment.

Jones: Do you have any advice for young people pursuing engineering careers?

Smith: They need to enjoy rigorous mathematical challenges. They also need patience because the scope of projects often changes in the middle. My most important advice is to pursue accounting and business while studying engineering. The best engineers understand both the technical and business aspects of a project. That is the bottom line, and most people forget that!

AN INTERVIEW between the Author and Deborah Wahl, Civil Engineer

Jones: As you know, this book is to help students in their career selection and specifically to help them decide about careers in engineering.

Wahl: Yes. How did the book come about?

Jones: It was a result of my teaching experience at Iowa State University. Because of my industrial experience, I was asked by the College of Engineering to participate in a freshman orientation program. I taught engineering design to students from all engineering fields. We spent a lot of time discussing many of the topics in this book. Tell me, Deborah, how you became interested in engineering.

Wahl: I became interested in engineering during my senior year of high school. I had always wanted to be an architect, but my uncle advised me to start in engineering so that I would not lose time if I decided to stay in engineering.

Jones: Why did he tell you that?

Wahl: He said that all freshman engineering courses would apply to an architecture degree, but that some freshman architecture courses may not satisfy the engineering college and that I might have to take engineering courses in subjects that I had already studied. This is because some engineering course requirements are more advanced than those of other fields.

Jones: An example would be in taking physics without calculus rather than with calculus. You might have to repeat physics.

Wahl: Yes.

Jones: Deborah, which engineering fields did you consider?

Wahl: Only civil. I did consider landscape architecture for a few weeks but decided that I wanted more of a challenge.

Jones: Did you go to Rutgers?

Wahl: Yes, Rutgers College of Engineering. I graduated in 1985.

Jones: What was your first job?

Wahl: I worked for a small company that did both civil engineering and land development work.

Jones: You designed structures—why are you laughing?

Wahl: Even now it seems amazing that I did that kind of work. I loved the field work, inspection of existing buildings. We crawled around in old hotels with flashlights and hard hats inspecting structures. It was like going back in time—seeing antique furniture and fixtures as if the doors were closed and time was stopped. I did a lot of inspection work and when I arrived on the job, the party meeting at the site would always say, "You're him?" [laughter]

Jones: They were expecting a man! After the inspection, what would you do?

Wahl: We would write a report and include an "as built" drawing. The existing structures were compared with local and federal codes, and new structures were designed to meet the codes. Cost estimates were made so that the renovations could be made.

Jones: I think I might like that job.

Wahl: We found a lot of strange things in those old buildings.

Jones: I can imagine. [laughter]

Wahl: We did a lot of new construction too. All of my designs had to be approved by the manager because I was new. I still have three years to wait before I can become a professional engineer.

Jones: You designed the house your family is building. Is there anything special about it?

Wahl: Just my bedroom. It has a loft with stairs and a skylight. [laughter] I get out there on weekends with my mother, brother, two sisters, and our boyfriends, and we help Dad pound the nails.

Jones: I know that you are working for a different company now. Why did you change companies?

Wahl: The owner of the first company decided to fulfill his lifetime dream to build a golf resort. He dropped his commitment to the company, and most all of us left.

Jones: What do you do now?

Wahl: I work for an engineering company in their land development division.

Jones: Shopping malls?

Wahl: I have been involved with a few. We handle the surveys, site plans, gradings and utility plans, drainage, sewers, water lines, roads and traffic patterns, and permits.

Jones: Do you survey?

Wahl: I would like to on a nice day. [laughter] We do a lot of waterfront work. [Ms. Wahl lives on the New Jersey shore.] We deal a lot with governmental agencies regarding permits, etc. We also do bridge inspections using boats, divers, and special truck-mounted platforms that can be lowered over the side of a bridge.

Jones: That must be fun in January!

Wahl: It does get windy.

Jones: Did your engineering degree prepare you for your work?

Wahl: No. I was interested only in site construction while I was in college, and I took too narrow a class load. When I got out of college I realized that I needed to know a lot more than I thought to do site works. I had to learn a lot of things on the outside that I could have at least obtained a basic background for in college. But I am not dissatisfied with the way it has turned out. I have been able to learn.

Jones: Most all of the interviewees in this book have said that learning to think and resolve problems systematically was more important than learning facts.

Wahl: That is right.

Jones: What advice do you have for students?

Wahl: If you are going to an engineering school, try to work at an engineering office during the summer or go on the co-op program. Try to get some work experience. I did not do this, but fortunately I love the work. Some of my fellow graduates did not like the particular type of work they were assigned to.

Jones: And that is the reason for the publication of this book. To help students obtain the information they need to make a career decision.

CONSTRUCTION ENGINEERING

A person engaged in construction engineering may be a civil

6.4. Here research is in progress on removal of organic material from waste water. Such research into purification of water provides valuable information for use in design of pollution control systems so that the environment will be further protected. (Photo courtesy of Phillips Petroleum Company)

engineer. However, there is a management-oriented curriculum designated *construction engineering*. Construction engineers in this category may take less basic science, mathematics, and engineering science than do civil engineers. Construction engineers have more time allowed in their curricula to study management in terms of human behavior, economics, law, and accounting. In other words, they are trained to supply managerial personnel to the construction industry.

The construction engineer is trained to start with a set of prints designed by civil engineers or architects and then to supervise the construction of the structures designated on the prints. For this reason, many construction engineers start their own companies after graduation. If you are planning a career in the building trade area, a degree in construction engineering would be a tremendous asset.

7 ELECTRICAL ENGINEERING

Do you have a high interest in electronics, electrical power distribution, communications, or computers? If you do, and if you have above-average ability to learn to utilize mathematics, then electrical engineering should provide you with an interesting and satisfying career. Electrical engineering is a large field. The Institute of Electrical and Electronics Engineers (IEEE) has over 150,000 members.

Electrical engineers need a good understanding of the physics of materials. They work with a large number of devices constructed of glass, paper, mica, plastics, metals, semiconducting materials (such as silicon and germanium), ceramic insulators (such as aluminum oxide, porcelain, and steatite), magnetic ceramics, organic liquids, and even gases.

The field of electrical engineering has always relied on the development of new electrical devices. When a new device is invented, it opens up many new technical areas. For that reason, device development is an important part of electrical engineering. The generator, motor, incandescent bulb, electron tube, transistor, and ceramic magnets have each had a great impact on electrical engineering.

Each electrical device, such as a transistor or ceramic magnet, has a distinct behavior under specified test conditions. However, when a number of devices are combined into a circuit, there are interactions that will determine the behavior of the circuit as a whole. It is the ability of electrical engineers to predict this circuit behavior that sets them apart from other engineers. It permits them to design circuits to perform specified functions.

Two of the oldest fields within electrical engineering are power distribution and communications. These two areas will continue to be important, and they will continue to hire large numbers of electrical engineers. For sheer size and increasing complexity, power distribution has to be one of the most challenging fields. The electrical engineer seldom designs a power system that is not interconnected with other systems. This means that system failures can be dramatic—especially in large cities.

There is something exciting and dramatic about seeing the "lights" come on when electricity is first supplied to an area from a new source such as hydroturbines located in a large dam. This is especially true in primitive areas that have not had electricity, but the residents of Arco, Idaho, had the same thrill in the fifties when the first electricity from nuclear energy "lit up" their town.

Telephone and radio communication have grown hand in hand. Electrical engineers find many exciting careers in communications from research to production and installation to general management. Satellite communication systems are a real bonus from the space program, permitting television transmission around the globe. Electrical engineers are working on many advanced communication systems for the future, including laser communication.

The space program as we know it today would not be possible without the technology of electrical engineering. Guidance and communication systems, monitoring devices, remote control systems, and environmental systems all have been developed with help of electrical engineers. The computer has been especially valuable in space exploration.

Computer engineering is a field of electrical engineering that has a dramatic impact on all of society. The computer is characterized by millions of circuit elements that have been previously developed to assure optimum performance and reliability. The development of the transistor to replace the bulky and less reliable electron tube is one of the dramatic achievements of this century. Large scale integration (LSI) has advanced computer technology more recently.

Electrical engineers are employed in all segments of industry and in many areas of government. The nature of their work is so varied that a degree of specialization is often desired even in the senior year of college. A student may want to plan a career in computer development, power, machinery, control, or any of a variety of areas.

7.1. Programming a robot. (Photo courtesy of Ford Motor Company)

Although there are many different electrical products and services available to industry, governmental institutions, and the general public today, you can rest assured that there will be many more in the future. If you want to see yourself as a benefactor of mankind through the application of electrical knowledge and theory, then electrical engineering is probably for you. You should learn more about this challenging field from electrical engineers working in your community.

Electrical engineering, a professional field since 1884, has expanded to become the largest technical profession in the world. Following is a list of the specialized fields of interest.

Acoustics, speech, and signal processing
Aerospace and electronic systems
Antennas and propagation
Broadcast, cable, and consumer electronics
Circuits and systems
Communications
Components, hybrids, and manufacturing technology
Computer
Consumer electronics
Control systems
Dielectrics and electrical insulation
Education
Electromagnetic compatibility
Electron devices
Engineering management
Engineering in medicine and biology
Geoscience and remote sensing
Industrial electronics
Industry applications
Information theory
Instrumentation and measurement
Lasers and electro-optics
Magnetics
Microwave theory and techniques
Nuclear and plasma sciences
Power engineering
Professional communication
Quantum electronics and applications
Reliability
Social implications of technology
Systems, manufacturing, and cybernetics
Ultrasonics, ferroelectronics, and frequency control
Vehicular technology

IEEE prints excellent employment guides for engineers and scientists, including student editions. These and other materials are available from IEEE or the Junior Engineering Technical Society (see Appendix D).

Watch for the advance of electrical engineering technology due to the introduction of the new ceramic superconductors!

AN INTERVIEW between the Author and Mark Schoenthal, Electrical Engineer

Jones: How did you become interested in engineering?

Schoenthal: This is a good question. I always had an interest in how things work and how to make them work.

Jones: What led you to electrical engineering?

Schoenthal: I guess it was an interest in radio, TV, and electronics.

Jones: You have a B.S. and an M.S. degree in electrical engineering from Drexel — right? — and also an M.S. degree in mathematics and computer science. Was your first engineering job at the Federal Aviation Administration [FAA] Technical Center?

Schoenthal: Yes. But to get through school, I worked in my parents' business (a fifty-room motel in Atlantic City, New Jersey) and also in many other tourist-oriented businesses each summer.

Jones: What is your current assignment at the FAA Technical Center?

Schoenthal: I am a systems engineer specializing in real-time software design. We currently are working with the Mode S [aircraft] beacon system. My work includes building the ARIES [aircraft reply and interference environment system] and the CID [communication interface drive—x-25 protocol simulation].

Jones: You are an amateur radio operator [N2AHS—author is KB2YA].

Schoenthal: Yes, as you know, I have helped to build a number of radio repeaters in the area.

Jones: What do you like most about electrical engineering?

Schoenthal: Assembling hardware components, writing the software, and getting the device to work. I also enjoy repairing your radios [the author's ham gear]. One interesting project we had was to build a mobile evaluation kit for aircraft transponders. We checked the U.S. fleet to see how well the transponders function.

Jones: Do you have any advice for young people pursuing an electrical engineering interest?

Schoenthal: Pick a good school and work hard!.

AN INTERVIEW between the Author and Mark Jankowski, Electrical Engineer

Jones: Mark, you are an engineer with both electrical engineering and mechanical engineering skills. How did that come about?

Jankowski: Quite by accident. When I was a senior in high school, I knew that I wanted to study engineering, but I didn't know which discipline. The University of Illinois offers a general engineering [GE] curriculum, so I enrolled and planned to change later. As I began my studies, I realized that the GE curriculum emphasizes systems engineering and design. I chose to remain in the GE department and select computer science as my secondary field.

Jones: What do you like most about engineering in general and electrical engineering in particular?

Jankowski: I enjoy the sense of accomplishment and pride after a successful project. I have worked primarily on automation projects in manufacturing plants. Watching a new piece of equipment that you designed make a person's job easier is a good feeling. Designing or installing a new control system that makes a plant financially competitive and saves jobs is even more rewarding.

Jones: How did your education prepare you for your work in industry?

Jankowski: I can honestly say that I think my education thoroughly prepared me to work in industry. The course work was important, but the problem-solving skills were even more important. Frequently, my job assignments are ambiguous and the stated goals are mutually exclusive. The same scientific approach to problem solving that I learned in the classroom also works in industry.

Jones: What advice do you have for young people considering engineering as a career?

Jankowski: Get involved immediately! I began my education as a very passive student. I was disillusioned with engineering until I started working in a manufacturing plant. I was impressed when I saw all the theory I was studying put to use. This gave me a new perspective on engineering. For me, engineering as a practical tool is much more exciting than engineering theory.

8 MECHANICAL-INDUSTRIAL ENGINEERING

Several fields of engineering traditionally concerned with the production of mechanical machines and devices have a broad impact on society. These fields are mechanical engineering, automotive engineering, aerospace engineering, and industrial engineering.

MECHANICAL ENGINEERING

Are you fascinated by the intricacies of complex mechanical machines and devices? Do you like to see what "makes them tick"? If you do, and if you enjoy physics and have ability to learn and apply mathematics, then you should investigate mechanical engineering.

The mechanical engineer is a generalist—a person with a broad background in engineering fundamentals—who adapts readily to a wide range of engineering situations. He or she understands the behavior of engineering materials under various conditions of force, heat, and pressure, and knows how heat is transferred in complex systems and how fluids and gases behave under a variety of circumstances.

As machine designers, mechanical engineers excel. They use basic knowledge of engineering fundaments in conjunction with a treasure chest of engineering skills developed by the various fields of engineering over the years to design internal combustion engines, gas

8.1. Savings in time and reduced scrap are gained from CAD/CAM (computer-aided design, and computer-aided manufacturing) techniques used to cut seat patterns for Chrysler cars and trucks. A computer-controlled trim cutter in Chrysler's Trim Plant (left) makes precision cuts in bolts of material according to CAD terminal directions (right) in the

turbines, control devices, appliances, jet engines, steam and nuclear power plants, earthmoving equipment, lathes and milling machines, diving equipment, air conditioning and heating systems, and literally thousands of other items. Just look around you. You probably can list a dozen products created by mechanical engineers — locks, hinges, switches, levers, tools, metal office equipment, and so forth.

Since mechanical engineering requires such a broad-based engineering education, over one-half of all firms employing engineers hire mechanical engineers. Mechanical engineers are working in almost all areas of industry — from the smallest engineering consulting firms to the very largest corporations. Job functions vary tremendously — from the intimate design of miniature devices to corporate management.

Mechanical engineers are often hired as test engineers. Test engineers evaluate the performance of systems, machines, or devices and make recommendations for improvements to mechanical engineers in research or development. Mechanical engineers may also be involved in production, sales, or management and other areas of

company's Corporate Engineering complex. Before adopting CAD/CAM, cutting seat patterns was a painstaking, manual job, requiring a worker to climb on top of the cloth and jigsaw paper patterns, cutting the cloth together with the pattern as one who sews does with a dress pattern. (Photo courtesy of Chrysler Corporation)

engineering. Ninety percent of all mechanical engineers work in industry.

More than 170 universities offer training in mechanical engineering. By all means, visit one near you if mechanical engineering appeals to you. Don't forget that mechanical engineering education can lead you into many different areas of specialization but that there are plenty of opportunities for those who want to do work of a broad and variable nature. Since one out of every four graduate engineers is a mechanical engineer, you should have little trouble finding a mechanical engineer to talk to in your community.

Over 10,000 new mechanical engineers are needed each year to work in energy, fusion, space, ocean technology, management, fossil fuel, cryogenics, transportation, manufacturing, and nonmachine systems. Computer simulation and control for manufacturing play a large role in engineering in general and in mechanical engineering specifically. About 400,000 men and women mechanical engineers are in the work force. Over 100,000 belong to The American Society of Mechanical Engineers. Write to them or to the Junior Engineering

Technical Society (see Appendix D) for more information.

AN INTERVIEW between the Author and Tom Gray, Mechanical Engineer

Jones: Tom, I think it is interesting how you obtained your education. Can you describe that?

Gray: Well, John, obtaining my education was a matter of economics more than anything else. When I graduated from high school, I had football scholarships to a number of small colleges, but none of them had good engineering schools. I only had enough money saved up to go to school for two years, so I decided that the best economical play would be to go to a two-year associate program, go out and start working, and then go back to night school to obtain my engineering degree. That is what I did. I received a two-year associate degree in manufacturing technology and then reenrolled in engineering and received a bachelor of mechanical engineering degree while attending at nights and then went back to get a master's in mechanical engineering and a master's in business management. It took me fifteen years to go through that whole cycle, but I think it was well worth it. The big thing you get by doing it that way is that you are working the whole time you are doing it, so you quickly factor out from your schooling what is really important and what isn't.

Jones: You were raised on a farm. Did working with farm machinery influence your decision to go into engineering?

Gray: No, I don't think being raised on a farm influenced my decision to go into engineering. I had left the farm by the time I was about ten years old and moved into the city. I think I didn't decide I wanted to go into engineering until probably my freshman, maybe sophomore, year in high school. It was like as always—at that age your big interests are in girls and cars, and I think probably working on cars and things like that made me decide I wanted to go into engineering.

Jones: You have an interest in electronics, and you are an amateur radio operator (KB8BM; author is KB2YA). Did you consider electrical engineering as well as mechanical engineering?

Gray: Yes, I have always had an interest in electronics and have been an amateur radio ham for many years. I was a ham even before I decided to go into engineering. When I was enrolling in engineering night school, I wanted to take electrical engineering, but, unfortunately, in the night school program there were only three areas of engineering available—civil engineering, mechanical engineering, and chemical engineering. I took my second option—I picked mechanical engineering—but what I did was I took all of the available technical electives I could in the electrical field, and I still try to keep up with

8.2. Test engineer "at freezing." Tests are being conducted under freezing conditions. (Photo courtesy of General Motors, Inc.)

all of the changes in electronics because nowadays mechanical and electrical engineering fields are so interwoven that you need a good understanding of both.

Jones: You have broad experience in a number of industries. What are the most interesting projects you have worked on?

Gray: I worked on many interesting projects, but probably the most challenging and interesting one was the building of a complete Greenfield factory for doing vinyl extrusion. What made it interesting and challenging was it had to be done from ground breaking to full production in a one-year period of time. It was a completely computer-controlled, state-of-the-art, automated factory for extrusion of vinyl products. That tight time frame and doing it on a controlled budget made it very challenging. It was one of the

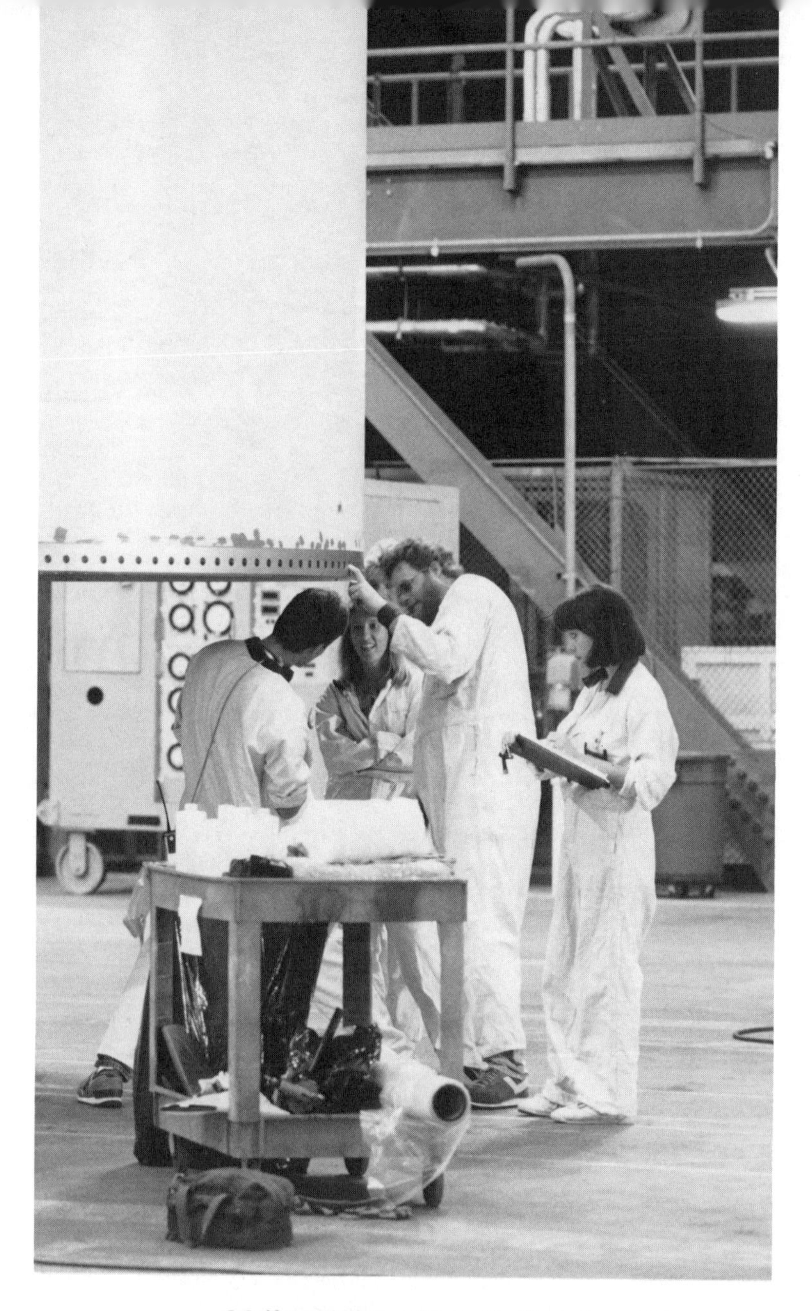

8.3. Kennedy Space Center, Florida. In the vehicle assembly building transfer aisle, technicians inspect a destacked solid rocket booster (SRB) segment. When the destack operation is complete, it will mark the conclusion of the disassembly of the Space Shuttle Atlantis. (Photo courtesy of NASA)

projects where you get a lot of gray hairs and you learn a lot. You also get a chance to apply all of the things you have learned from all of the other projects into one major project.

Jones: How has your education supported your career as an engineer and as an engineering manager?

Gray: I would have to say that the technical portion of my education—engineering courses, those kinds of things—supported my career as an engineer and as an engineering manager probably fairly well. What is lacking in engineering education—I think this is probably pretty universal through all of the engineering schools—is that they do not do a good job of educating young engineers in interpersonal skills, management skills, and the people-type skills, and, as a result, many engineers that find their way into the management ranks are poorly prepared for the jobs that they have found themselves in. I would suggest that engineers who want to become managers reinforce their education with additional course work, seminars and workshops, and things like that in the area of people management.

Jones: What advice do you have for young people considering a career in engineering?

Gray: That is a good question: What advice would I give young people who are considering a career in engineering? The problem you have here, I think, is that most people except other engineers don't really understand or know what engineers do. I would suggest that someone considering engineering as a career who didn't have a good understanding first decide what type of engineering work she or he would like to do, whether it be electrical or mechanical or computer science or civil or any of the disciplines of engineering. Once you have decided what discipline you would like to pursue, I would suggest you seek out at least ten engineers working in that field and interview them about what they actually do—what is their job function—and I would also make sure that the ten people you interview are from ten different industries. That way it will give you a broader perspective of what people do in the career that you are going to choose for your future.

AN INTERVIEW between the Author and Bill Beach, Mechanical Engineer

Jones: When did you become interested in engineering?

Beach: In high school in 1963 or 1964, although I was more interested in natural science at that time.

Jones: That's interesting. I was also interested in natural science in junior and high school. Why did you decide to go into engineering?

Beach: I started in natural science, but I needed a way to work my way through college so I switched to engineering.

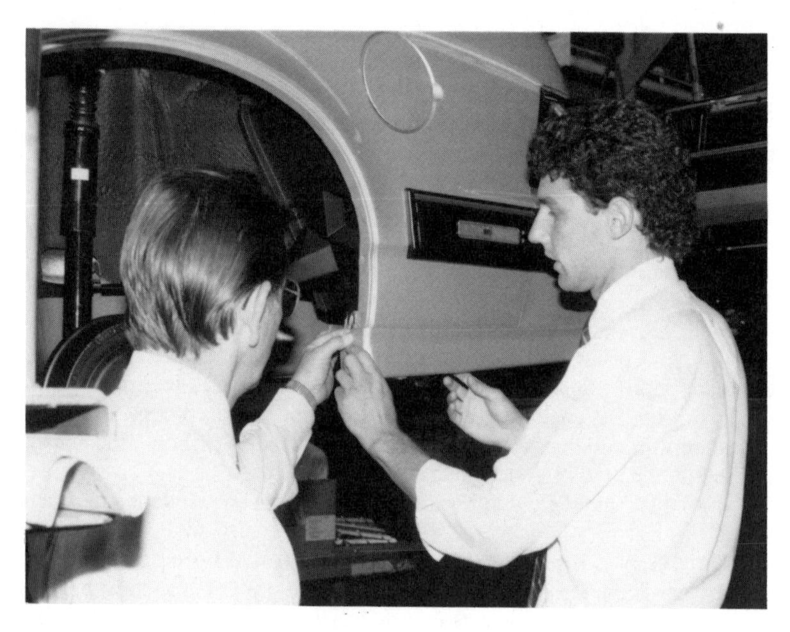

8.4. Quality is "the game." (Photo courtesy of General Motors, Inc.)

Jones: How did that help finance your education?

Beach: I could go to school on a co-op program.

Jones: For our young readers, what is a co-op?

Beach: You work for a company for one academic period and go to school the next, and so forth. My counterpart had the same job but during alternate periods.

Jones: Who did you co-op with?

Beach: The Red Stone Arsenal. I worked with a survey team at the small missile test ground and then at their aeronautical laboratory.

Jones: Why did you choose mechanical engineering?

Beach: Because I like to design machines.

Jones: What engineering school did you go to?

Beach: University of Alabama. I left the co-op system after two years and worked full-time as an engineering draftsman. I finished school at night.

Jones: After graduation, what?

Beach: I stayed with my company for six months and then joined Owen Corning Fiberglass. I worked in a factory as a mechanical project engineer.

Jones: What kind of projects did you handle?

Beach: Machine design and also modified high-speed winders for the glass fibers.

8.5. Design, design, design. (Photo courtesy of General Motors, Inc.)

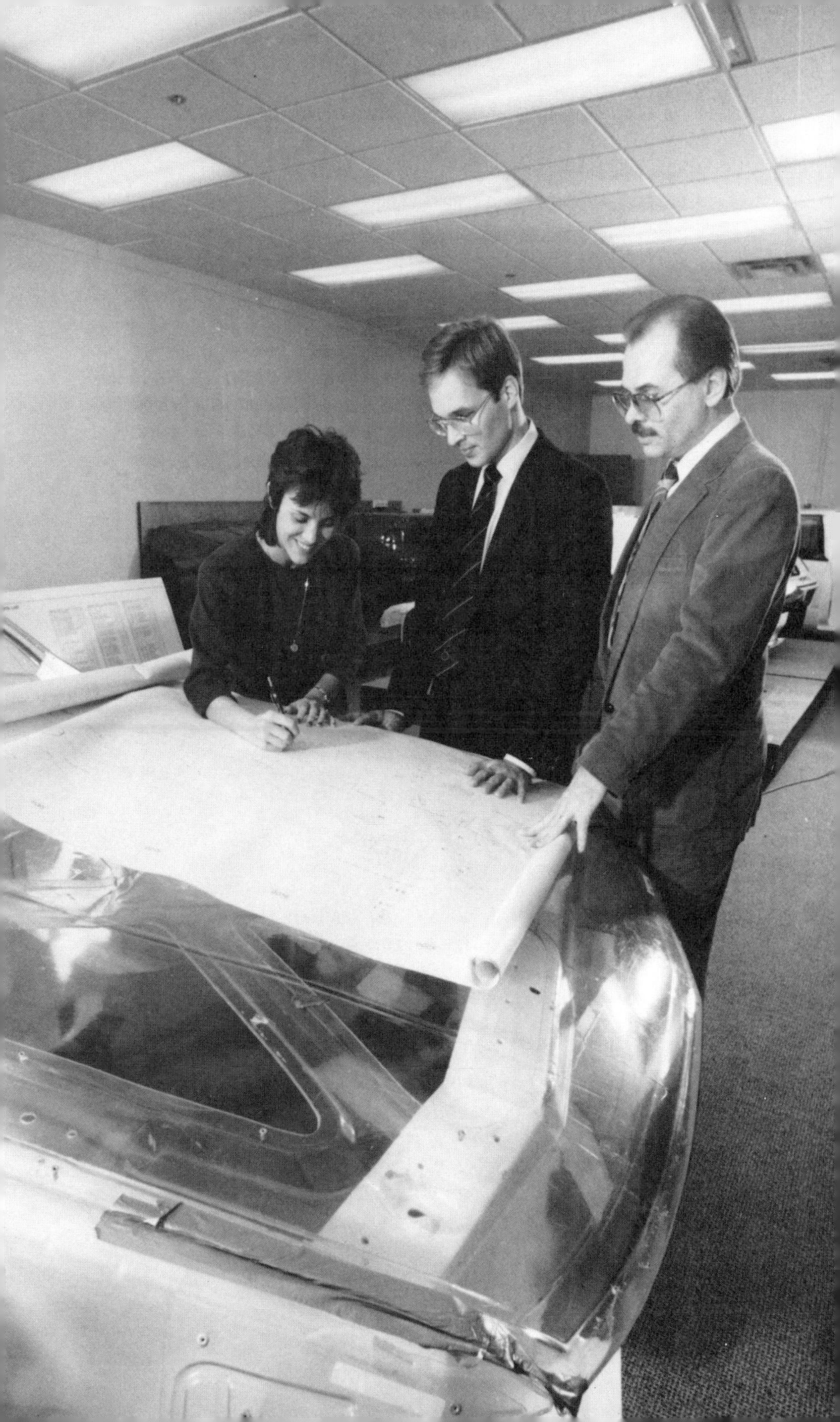

Jones: How long were you there?

Beach: Two and one-half years. They had a layoff and I got laid off. [laughter]

Jones: And then where did you go?

Beach: Fieldcrest Mills in Eden, North Carolina. I worked as a project engineer on the corporate engineering staff. The projects were large in scope, mainly in mechanical equipment installation. I was there three years. After that, I went to Anderson, South Carolina, and started an engineering consulting business that included machine design and structural work. I then went to American Thread Company in Marion, North Carolina, to become manager of mechanical engineering. I also supervised civil and structural engineers. Later, I went to a machinery manufacturing company that folded, and I worked for a consulting engineering company until I took my present job.

Jones: You have moved around quite a bit. Has that been good or bad?

Beach: Good, because I was able to learn more than one industry and technical field.

Jones: What do you like best about engineering?

Beach: The variety of work. It changes from week to week. I like to do analysis work best. It is something you can do by yourself. Unfortunately, analysis pays the least. I also like to supervise small engineering groups.

Jones: Any advice for young people?

Beach: Yes, gear yourself to becoming a manager. No matter where you go, you will need management skills. This includes courses like public speaking. You must develop communication skills as well as technical skills.

AUTOMOTIVE ENGINEERING

Presently there are no universities training students specifically for the automotive engineering degree. If you have a specific interest in this area and do not live near a university with an engineering program geared toward the automotive industry, you probably can still easily get into the automotive industry with a mechanical engineering degree (or also with most any engineering degree). Many universities have Society for Automotive Engineers (SAE) chapters (yes the same "SAE" as on your oil can).

The SAE has 52,000 members (1988) in fifty-two countries. Members include mechanical, electrical, aeronautical, metallurgical, chemical, industrial, civil, ceramic, and computer engineers. The society is currently focusing on space exploration, so that the advancements in space will be based on standards developed by SAE.

As you can see, SAE is dedicated to the advancement of mobility in any environment—land, sea, air, or space. You can learn more by talking to engineers in this field or by writing to the sources in Appendix D.

AEROSPACE ENGINEERING

If you have an interest in things mechanical, are curious about the frontiers of air and space travel, or have a secret desire to be an astronaut—and the ability to obtain the desire—you should consider aerospace engineering. The aerospace industry is one of the largest in the country. Aerospace engineers will be in continued demand as new space programs and commercial aircraft develop.

The research and development tasks of the aerospace industry are consistently in the vanguard of the latest technology applications. Aerospace engineers must, therefore, stay aware of the newest developments in materials and processes, electronics, and science so that these, after having been carefully considered and tested, may be applied at the earliest possible time in the design and production of aerospace vehicles and programs. This is necessary to meet the competition that is an earmark of the industry. Such engineers are sometimes required to work under high stress situations to meet proposal, contract, and production deadlines. However, these challenges, with the requirement of the continued learning needed to meet them, are intellectually appealing to many. In aerospace engineering, there are opportunities for both intellectual and financial rewards.

Aerospace engineers have good technical backgrounds and many work in areas of industry other than the aerospace industry. In fact, thousands of space program developments have been commercialized by many different industries.

This is an age of flight, both in the atmosphere and in space. It seems certain that governments will have to continue to spend substantial quantities of money for commercial developments, scientific projects, and defense. Many technological advances will be made that will enhance the careers of aerospace engineers in government and industry. If you have the desire for aerospace engineering, go after it!

8.6. The DC-10 is a multi-range trijet transport designed to answer airline needs in the 1980s and beyond. It can carry up to 380 passengers in four commercial versions for economical operations on route segments from 300 to more than 6,000 statute miles (483 to 9,656 km). (Photo courtesy of McDonnell Douglas, Inc.)

8.7. The McDonnell Douglas MD-80 family of twin-engine jets helps solve the environmental and economic problems facing airline operators with a blend of new and mature technology, while continuing the highest standards of passenger comfort and efficient cargo transportation over short-to-medium–range routes. (Photo courtesy of McDonnell Douglas, Inc.)

8.8. The F/A-18 Hornet is a high-performance tactical airplane that can operate from either aircraft carriers or land bases. It is a single-seat, twin-engine aircraft which can undertake fighter, attack, and reconnaissance missions. (Photo courtesy of McDonnell Douglas, Inc.)

8.9. The F-15 was designed to outperform and outfight any enemy fighter aircraft. The fixed wing, twin-engine aircraft has a primary mission of attaining air superiority. The Eagle also has an excellent day air-to-ground weapon delivery capability. The F-15 combines an advanced fire control system with multiple air-to-air weapons for optimum combat efficiency. It can carry conventional air-ground ordinance without off-loading any of its air-to-air missiles. (Photo courtesy of McDonnell Douglas, Inc.)

INDUSTRIAL ENGINEERING

If you are interested in the relationships between people and their work, you may want to become an industrial engineer. Industrial engineers are concerned with the relations between people, machines, energy, materials, and money. Industrial engineers select operating processes and methods to accomplish various tasks. These could be such things as methods for manufacturing an electrical appliance or a system to determine the cost of a particular office procedure. Variety is the key word in industrial engineering.

Industrial engineers develop standards for evaluating the performance of workers doing various tasks. These evaluations can be used to determine proper incentives and compensations for the workers. The working conditions are modified to increase productivity by improving surroundings, removing safety hazards, and eliminating excessive labor and lost motion. The morale of the workers is an important consideration at all times.

The movement of materials through industrial operations is a prime consideration in industrial engineering. Plants are located and designed to assure maximum efficiency in material movement. Industrial engineers, often working with engineers from other fields, design the layout of machinery and production lines using scale models or blocks to represent machines and other equipment. The plant layout frequently determines the type and size of building for new operations.

Measurement and control systems are established by industrial engineers to assure quality in manufactured products. Statistical methods are used to follow the efficiency of a particular operation. The availability and cost of materials is continuously monitored to prevent shutdown due to the temporary exhaustion of a crucial raw material or due to underpricing of the final product.

Industrial engineers study the overall operations of manufacturing companies, governmental agencies, banks, and other such organizations. The results are reduced into systematic studies that predict the overall capability of each organization to handle money, people, energy, equipment, and materials. These studies may result in the better use of capital, improved employee morale, more potential for expansion, and other benefits.

Some students in college prefer a curriculum in engineering operations (usually taught in an industrial engineering department), which allows either a broad background in business, economics, and management or a special emphasis of one area of industrial engineering. The student has more freedom and usually takes fewer engineering courses than do most industrial engineers.

Industrial engineering is an expanding field. There are plenty of jobs, many of which lead to management. Industrial engineers work in all parts of government and industry, which allows them to choose the localities in which they work. If you like to work with people, industrial engineering may be for you.

This is one of the most exciting times for industrial engineering. Why? Computers! Computer simulation of proposed manufacturing operations, computer control of operations, and computer design of operations are applied not only to industry but also to banks, hospitals, and almost every walk of life. Because of these and numerous other technologies, industrial engineering must continually change to help mastermind adjustments for industry and other organizations.

There are well over 100,000 industrial engineers at work, and the demand for them will continue to increase. You can be a "stay-at-home" and probably find work, if that is your desire. You will have the largest job market. Because of the wide job choice, you will have to weed out companies that do not use the skills of industrial engineers efficiently. You can do this by asking specific questions to determine exactly what a company's industrial engineers are doing and what role they are playing in management. For more information, write to the Institute of Industrial Engineers or the Junior Engineering Technical Society (see Appendix D).

AN INTERVIEW between the Author and Bruce Guthrie, Industrial Engineer

Jones: Bruce, you went to the University of Rhode Island and graduated in industrial engineering in 1970. How did you first learn about industrial engineering?

Guthrie: I started in chemical engineering but transferred into industrial engineering after my first year in college. I was interested in the interpersonal relation in industrial engineering. I was interested in methods and labor more than just the process.

Jones: What types of work have you done as an industrial engineer, and what was the most enjoyable?

Guthrie: The most enjoyable is starting out with a problem I have defined, suggesting a program to management, finding the solutions, justifying the approach, starting and completing the project, and then judging its success. I have worked as a division consultant for a large company, but that is not as enjoyable to me as handling my own projects.

Jones: How did your education prepare you for industry?

Guthrie: I have a B.S. in industrial engineering, and there I "learned to learn." I learned to find information and to solve problems. I also have an M.A. degree in management, which gives me a broader view of corporate operations. And I have an M.A. in personnel management, which is valuable as I work with people as an industrial engineer.

Jones: Do you have advice for young people who may pursue a career in industrial engineering?

Guthrie: In high school, try to get a good background in mathematics and science. You must also develop writing skills to sell an approach to an engineering problem to management. On your first job, learn to listen and absorb all the information you can to establish credibility. Learn to approach a problem from all points of view to develop the best solution. It may not be the lowest- or highest-cost solution, but it should be the most workable.

9 AGRICULTURAL ENGINEERING

Engineering is an important part of the world's largest industry, agriculture. Agricultural engineers design and improve farm machinery; construct crop storage and livestock buildings; develop drainage and irrigation systems; design chemical-application equipment for production improvement and pest and disease control; and design specific cropping, water control, and waste disposal systems. From this list, you can see that agricultural engineers do more outdoor work than other engineers do.

Agricultural engineers work as farmers or contractors or work in consulting. Large farm machinery corporations; food processing firms; and manufacturers of irrigation equipment, animal confinement systems, and pumps and hydraulic equipment hire many "ag" engineers. Numerous smaller industries manufacturing agricultural equipment also employ "ag e's."

Federal, state, and county governments hire agricultural engineers. The Soil Conservation Service, Corps of Engineers, U.S. Geological Survey, Bureau of Land Management, and Bureau of Reclamation all hire agricultural engineers. In addition, agricultural engineers find employment with such groups as pollution control agencies, resource councils, and recreational agencies.

All countries in the world are trying to improve their agricultural practices. For this reason, agricultural engineers have a good opportunity to travel around the globe. Many of these opportunities are offered by universities, the United Nations, and federal agencies.

There is one other advantage to agricultural engineering—the opportunity to work with the best people in the world, the farmers.

Technology areas defined by the American Society of Agricultural Engineers are power and machinery, soil and water, structures and environment, electrical and electronic systems, and food and processing engineering. Forestry, aquaculture (such as growing oysters), food management, and biomass applications (such as producing fuel from sunflower seeds) are important work areas.

Because I am allergic to farm work, I am all for the development of a remote-controlled tractor. Can't you see yourself directing it from your front porch as it plows the field and you sip lemonade?

AN INTERVIEW between the author and Mike Asche, Agricultural Engineer

Jones: Mike, you graduated from Clemson in 1976.

Asche: Yes, a beautiful school.

Jones: Why do you say "beautiful school?"

Asche: They had an undergraduate capacity of ten thousand students at that time. There were only a few students in agricultural engineering and I enjoyed both the special attention given in small classes and the people.

Jones: What led you to agricultural engineering?

Asche: A lot of things. I started in premedicine, but organic chemistry turned me off because of the memorization. I like mathematics and problem solving, and, after looking at mechanical and civil engineering and seeing that these fields were applied to agriculture, I decided agricultural engineering was for me. I was raised in an agricultural environment. My dad had a feed mill and I was associated with farming. I did special crop spraying to work my way through college.

Jones: Fortunately, organic chemistry is taught differently now, but I do remember the memorization. I understand that you have the same job you started with.

Asche: Well, I am with the same company, John Deere, but I have had five different jobs. I started as a field service manager in New York State. I did that for four years. I trained technicians in how to repair farm equipment. I did a lot of public speaking and teaching. I was promoted to product specialist. This was because of my "art of B. S." [laughter] I covered thirteen states and put on meetings and demonstrations for customers and dealers. That job was very public-speaking oriented because that was all I did!

Jones: Then what happened to you?

Asche: The main field job at Deere is territory manager, which is what I do now. I am the go-between connecting the dealer and John Deere. I get the product to the dealer and follow up on business details and give any other services the dealer needs. We went from 62,000 employees in 1981 to 33,000–34,000 today, and we territory managers feel we have absorbed all the extra work. [laughter] I think this is a sign of the economic times.

Jones: I think you are right.

Asche: My current territory [Delmarva Peninsula plus part of New Jersey] was an increase in territory size that also allowed me to work near my home.

Jones: John Deere has been good to you.

Asche: Yes, they have! I had fourteen job offers and was considered for mechanical engineering and civil engineering jobs as well as agricultural engineering. I liked the people at John Deere.

Jones: I don't think I have to ask you if you would do it all over again.

Asche: I am very comfortable at John Deere. They have brought my management skills up to that of an M.B.A. person. I would have to say yes.

Jones: Do you have advice for students in high school and college?

Asche: Yes, they should take as much accounting, economics, and other business courses as possible. They should broaden their horizons and remember that they will probably be working in business some day.

Jones: How has your education helped you?

Asche: Engineering gives you the logical methods of problem solving, the systematic approach to accomplishing something.

PART THREE

SHOULD YOU BE AN ENGINEER?

In this section, criteria are established to help a person decide if she or he has the personality of a potential engineer and, if so, what type of engineer?

A person's educational background is also important, as is one's aptitude for a particular engineering field. These aspects will be discussed. Then details as to how a person should pursue career goals are given. (See also Appendixes A, B, D, and E.)

10 THE ENGINEER'S PERSONALITY

Engineers are people. They are a cross section of the more intelligent or better-educated portion of society, and therefore the engineering field includes many personality types. However, either engineering propagates certain personality types or certain types are more likely to become engineers.

Lee Harrisberger, professor and department head of mechanical, industrial, and nuclear engineering at the University of Cincinnati, has summarized studies made on the traits of engineers that were published in *Machine Design* in 1959 and 1960. His book *Engineersmanship —A Philosophy of Design* copyright © PWS-KENT Publishing Co. is a delight to read. His summary is printed here by permission.

PORTRAIT OF A TYPICAL ENGINEER

What is a typical engineer? What kind of person becomes one? Raudsepp correlated the results of a large group of studies made by psychologists to determine the personality, habits, attitudes, and abilities of the engineer. Here is what the typical engineer looks like:

Personally:
Reserved, self-sufficient, independent, well adjusted.
Has little interest in people — is object and idea oriented; tends

toward shyness and is not emotional or impulsive.
Industrious, determined; serious, sincere, honest, orderly.

Intellectually:
Very intelligent — in the upper 10 percent of the population — prefers math and science.
Tends to be narrow; has few cultural interests; reads little outside field.
High mechanical and visual ability but is poor at communication.

Socially:
Not gregarious; has a casual relationship with colleagues.
Superficially friendly; avoids leadership in group activities.
Insensitive to needs of others; has few intimate friends.
Tends to conform socially; dislikes chit-chat.

At work:
Action oriented, hard worker, eager beaver, well organized.
Highly motivated to achieve a successful solution.
Prefers concrete, orderly mechanical tasks; tends to be conventional.
Works hard to avoid criticism and failure; cautious, and conservative.
Respects authority but dislikes being supervised.

At home:
Excellent spouse and parent — family oriented.
Solid middle-class suburbanite; willing worker in community projects; interested in gardening, home repair, and crafts.

At play:
Active hobbyist; primarily mechanical hobbies; strong interest in outdoor life.
Spectator of team sports; active participant in non-competitive sport activities. Willing participant in family-oriented recreation.

To sum up:
An intelligent, hard-working, independent, action-oriented, solid citizen and family person.

Of course, this stereotypical engineer doesn't really exist. The profile is based on a large number of individuals with a job title of "engineer." Many of these are not graduate engineers. However, you would find a number of these listed traits in many engineers. Reading this profile, you would think that engineers at work are similar to machines. Now, that just isn't true! Most of the engineers I know are humorous, intelligent, good-looking, charming, wonderful people!

The portrait of a design engineer (or development engineer) as summarized by Professor Harrisberger follows.

PORTRAIT OF A DESIGN ENGINEER

What contrasts the design engineer from other engineers? What are the attributes of an engineer that makes her or him a good design engineer? Raudsepp also surveyed the attributes of the creative design engineer. Here is what the typical design engineer looks like:

Personally:
Very self-confident; willing to take calculated risks; open-minded to experience; highly tolerant of criticism; distrusts routine; dislikes regimentation, has strong initiative; uninhibited in communication of ideas; is a constructive nonconformist; imaginative.

Intellectually:
Strong background in fundamentals, with broad interests; very curious and observant; highly creative; persistent and patient in pursuit of a solution; looks for alternatives and ambiguities; very flexible; tolerant of ideas and innovations; has strong interests in mechanical, scientific, artistic, and literary areas.

To sum up:
In contrast to other engineers, takes the creative nonconventional approach; is willing to take risks; is not afraid of failure; is idea-oriented and creative; and has a much broader intellectual scope.

You can see the contrast between these two personality types. If you don't see yourself in these portraits, and even if others agree with you that you do not fit, it doesn't mean you will not be satisfied with a career in engineering. All normal personality types are working in engineering, and all types are successful and satisfied.

If you studied the activities of engineers on an individual basis, you would find a tremendous variety of interests. While some engineers have a strong mechanical interest and like to repair their own cars and household appliances, others would rather do artwork or study music or take part in theatrical activities. Many engineers have a strong interest in the biological sciences, and you might find them looking for edible wild plants or studying birds on the weekend.

Engineers are engaged in a wide range of community activities. Some are Boy Scout leaders, church leaders, or members of fraternal organizations such as Lions, Elks, or Jaycees. They may hold local political offices or be engaged in fund-raising for the United Fund or cancer research.

Engineering can lead into biomedical research, medicine, forestry, military service, commercial flying, governmental service, consulting, technical writing for magazines or newspapers, merchandising, buying, law, teaching in colleges or technical schools, the sciences (I know one electrical engineering graduate who is a professor of physics), and many other vocational areas. If aptitude tests show that you have a particular interest in some vocational area, check further to see if engineering training of the proper type would enhance your career.

11 PREPARING FOR COLLEGE

By now you probably have the impression that all engineers are mathematical geniuses, well versed in all of the sciences from agronomy to zoology. At the twinkle of an eye they can design a locomotive or a stop watch. You may also think that all engineering programs and schools are very difficult to get into (and out of). Actually, guidance is given to students on an individual basis to help them adjust to college work in engineering. The design and analytical skills come from persistent hard work.

It is true that it is desirable for a beginning engineering student to have a high school background and interest in mathematics, chemistry or physics (preferably both), and English. In fact, it is an advantage to take all of the mathematics offered in high school. But if you find that you are about ready to graduate from high school and do not have the desired background for engineering, it doesn't mean you cannot go into engineering. It does mean that you will have to spend more time or energy in college to make up your shortcoming. I decided to go into engineering after finishing military service plus two years of college. However, it is best by far to make a decision to study engineering early.

English is a difficult subject for some engineering students, but its importance can not be overemphasized. Engineers are required to report their progress and achievements in a logical and precise manner. In addition, they frequently are required to instruct others who are engaged in engineering projects. Many engineering devices cannot be constructed or assembled without proper written instructions from the design engineer. Engineers also must convey technical information to the public in terminology the public understands.

If you are in high school, a close relationship with your guidance counselor can help you obtain information from universities about engineering educational programs in your state as well as other parts of the country. You will be able to determine entrance requirements to educational institutions and also the availability of scholarships, housing, and part-time employment opportunities. Most guidance counselors also have information on the cost of tuition, books, and living (room and board).

JETS (Junior Engineering Technical Society, 1420 King Street, Alexandria, VA 22314-2715) for thirty-five years has maintained a focus on academic excellence. According to JETS informational materials, JETS is the only well-established nationwide high school program that focuses on the application of science and mathematics to engineering and technology. JETS materials can be used in existing high school clubs and classes. This enables and encourages you to associate with other students interested in engineering and technology.

JETS sponsors the National Engineering Aptitude Search (NEAS), which is a test of mathematical ability, higher-order thinking skills, and problem-solving skills. Interests of the student as related to engineering are assessed. The results of this testing can help a student find his or her place in engineering or technology. The results might also show areas of needed improvement that can be corrected by study of specific subjects. The NEAS test is rigorous; it is designed to compare only students who are going into engineering and not all high school students. The NEAS is conducted at colleges and universities only. JETS will send you a booklet giving the time and location of such testing.

JETS also sponsors TEAMS (Tests of Engineering Aptitude, Mathematics and Science) competition, which allows academic competition in mathematics and science between high schools. More information can be obtained directly from JETS. If you participate in NEAS or TEAMS, you will receive a *JETS Report* newsletter subscription plus a guidance mailing at your home.

The cost of education is increasing, and parents are often hard-pressed to meet all of the needs of a college student. Therefore, prospective engineering students should make some effort to earn and save money prior to beginning college. This is good advice even to those of you whose parents are well-heeled or perhaps have available funds due to some other set of circumstances. If you earn part of the money you need for college yourself, you will learn the value of money and will be able to say also that you worked your way through college. Many potential employers will appreciate that.

Military service should not be counted out as a source of college financing. Not only that, valuable educational opportunities are offered in the service with excellent instruction in complex areas often related to engineering. The advertising you see on T.V. is not all hype. ROTC, which I enjoyed, is another possible source of college funds.

I served in the Korean War in the 17th Regimental Combat Team, 7th Division. But before I went to Korea, I spent twenty-two weeks at the Army Artillery School, Fort Sill, Oklahoma and learned mathematics, surveying, sound ranging, communications, and air transportability taught by the best instructor I have seen anywhere. (So, I ended up in the infantry! The Army can get confused at times.) After spending time in the service, I received the GI Bill benefits, which got me through college with a little (a lot of) part-time work.

At the time this book is being written, the Montgomery GI Bill provides for $10,800 to be applied toward education in return for three years of military service. The services also have a 75–90 percent tuition refund program for military personnel attending college courses. And there is even an eleven-month program that helps prepare people for college ROTC or service academies. It is called "BOOST," and its purpose is to prepare students to take the SAT (Scholastic Aptitude Test). There is no military service commitment required upon completion of this program, but those who score highest on the SAT can receive $25,200 from the Army or Navy College Fund.

It is useful to know that there are *engineering technology* programs related to almost all fields of engineering. The technologist in *mechanical technology,* for example, would have employment that lies between that of the mechanical engineer and the machinist or mechanic. However, he would be closer to the engineer. Engineering technology programs are different from engineering programs and should be investigated through your counselor.

One of the best ways to prepare yourself for college is to learn to study properly. Your high school teachers can help you tremendously in this area if you request the help. Learn to schedule your time, select proper studying conditions, and read and write properly. You should learn how to organize your notes from class and from reading. When you work problems, be neat, complete, and logical in your approach. You can't imagine how much your college professors will appreciate that.

I highly recommend that you try to associate with men and women you know who are engineers. They can help you tremendously—there is even an outside chance that they may get you a scholarship.

12 YOUR ENGINEERING PREFERENCE

By now you are probably leaning toward one or two fields of engineering more than others. If you prefer to manufacture chemicals rather than fabricate or create products from them, you may be more interested in chemical engineering than in ceramic engineering, metallurgical engineering, or polymer engineering. The same would be true if you were more interested in processing than in physical properties of the final products.

If you prefer the mechanical fields but have more interest in production methods than in machine design, you will probably investigate industrial engineering more strongly than mechanical. But if you desire a stronger background in engineering fundamentals before pursuing a career in industrial management, you may choose mechanical engineering.

If you plan a career in space technology you may prefer electronics over power engineering. A very artistic person may prefer architecture over construction engineering. And a person with a deep interest in agriculture may prefer agricultural engineering over civil or mechanical engineering. However, if a person wanted only to design farm machinery, she or he may prefer the stronger design background of mechanical engineering.

To help make a decision on which engineering field you choose, you should ask yourself some questions and then evaluate the answers. Typical questions might be

1. What are my vocational goals, and which of the engineering fields would most likely lead me to achieving them?

2. In which of these fields is it within my ability to do satisfactory work?

3. Which of these fields would most likely supply me with employment opportunities within the geographical area I prefer?

4. Which field would give me the broadest possible education and yet train me in the area of my vocational preference?

5. Which of these fields are taught in a university that I can be admitted to and can afford to attend?

After you list your questions, you may find that you do not know the complete answers. Then it is time to talk to engineers working in your areas of preference and to visit engineering schools. You can do a lot by correspondence, writing directly to engineers and engineering colleges. Your high school vocational counselor may have materials on hand that you can read and may be able to direct you to others.

It is often difficult to make a career decision. Some college students (including me) have waited into their second year before deciding for which vocational field they should prepare. The purpose of this book, of course, is to help high school students and even first-year college students make early, intelligent decisions about engineering careers.

Five appendixes that may be helpful to you follow this chapter:

Appendixes A and B list a number of engineering schools. There may be other schools in your area. Note that some schools have evening programs or off-campus programs. Check those in your area for up-to-date information. Write to the schools for detailed information on curricula. The lists in Appendixes A and B were prepared by the Accreditation Board for Engineering and Engineering Technology and were updated in 1989.

Appendix C lists typical curricula for various engineering fields. Because the freshman year varies little, only the last three years are listed.

Appendix D lists the engineering societies. You can write directly to them for information in addition to what you can get from the engineering schools. Write to the schools of interest for current curricula.

Appendix E is a list of selected references you may be able to obtain.

For those who may be interested in engineering technology that allows career entry after a two-year program leading to an associate degree, ABET-approved programs are listed in Appendix B. Four-year programs leading to a B.S. degree are also listed.

During your investigation of engineering fields, I suggest that you try to get more than one view from people working in each field you are investigating. When you are finished with your data collection, make a list of the fields you have investigated and rate them according to how satisfied you are with them based on the answers to your questions. Then make your selection!

EPILOGUE

A young man came into my office requesting information on piezoelectric devices—crystalline materials that will change dimensions when an alternating electrical field is applied (the reverse effect is also possible). These materials are used in ultrasonic cleaners, sonar equipment for underwater detection, and radio transmission. He had learned that high voltages (twenty thousand volts) could be obtained from piezoelectric devices and were being used to start small engines such as in lawn mowers.

This young fellow was in the process of designing a home that would utilize solar energy as a source of heating and electricity. He seemed to have a good knowledge in certain areas, and I asked him if he was an engineering student. He said that he had been employed for two years (after military service) as a shipping clerk. He had one year of junior college and two years of university training in geology.

I found on further discussion that he had not studied engineering because of false impressions about engineering he had had in his youth. He thought that the engineering curriculum was "stifling." Actually, most engineering students enjoy their programs. He could not see (until I told him) that he had the rare attributes of the design engineer. He had broad interest and was very intelligent. He just had to know how things work. He didn't like to be restricted to routine tasks.

If he had studied for four years in an accredited engineering curriculum, he would now be a very successful design engineer rather than a shipping clerk. As it was, he had to seek advice on every engineering point when he tried to design. He had to do the thing he loves most on his own time—for no pay.

There was still a chance for him because he was not tied down by debt. He was young enough to go back to school for a couple of years and earn a degree that would mean more to him. If you are this type of person, don't waste your time and money on education that will not advance your career. Find out what the facts really are, and I am sure that you will find that engineering is for you.

APPENDIX A

Accredited Engineering Programs in the United States

Part II-A
Accredited Programs Leading to Bachelor's Degrees in Engineering, 1989 by Program Area

Note: The listings below are grouped according to the title of the program as reported by the institution and accredited by the Engineering Accreditation Commission (EAC) of the Accreditation Board for Engineering and Technology (ABET). Options under overall program titles are listed with the major programs having the same title as the option, and cross-referenced in brackets under the title of the overall program, where appropriate. The letters in parentheses after the program group headings indicate the Participating Bodies of ABET that have been assigned curricular responsibility for assisting ABET in the preparation of program criteria and the appointment of program evaluators for programs within their areas of professional disciplinary competence. (See list following this tabulation.) Lead societies have primary but not exclusive responsibility for assigned program areas.

This listing includes only baccalaureate degree programs. For master's degree programs, refer to the separate listing that follows. Users should note that similar or related programs may be listed under variant titles.

AEROSPACE GROUP (AIAA)

(Programs in this group are accredited according to the program criteria for Aerospace and similarly named engineering programs.)

Aeronautical and Astronautical Engineering

Illinois at Urbana-Champaign, University of bdC
Ohio State University bd
Purdue University, West Lafayette bdC

Aeronautical Engineering

California Polytechnic State University, San Luis Obispo bdC
Embry-Riddle Aeronautical University, Daytona Beach Campus bd
Embry-Riddle Aeronautical University, Prescott, Arizona Campus bd
Rensselaer Polytechnic Institute bdC
United States Air Force Academy bd
Wichita State University bde

Aeronautical Science and Engineering

California, Davis; University of bd

Aeronautics and Astronautics

Massachusetts Institute of Technology bdC
Washington, University of bd

Aerospace and Ocean Engineering[1]

Virginia Polytechnic Institute and State University bdC

[1]Joint with SNAME under Ocean Group
[2]Joint with ASME under Mechanical Group

Aerospace Engineering

Alabama, University of bdC
Arizona State University bd
Arizona, University of bd
Auburn University bdC
Boston University bd
California State Polytechnic University, Pomona bd
California, Los Angeles; University of bd
Central Florida, University of bd
Cincinnati, University of bC
Florida, University of bdC
Georgia Institute of Technology bdeC
Illinois Institute of Technology bdC
Iowa State University bdC
Kansas, University of bd
Maryland, University of bdC
Michigan, University of; Ann Arbor bd
Mississippi State University bdC
Missouri-Rolla, University of bd
New York at Buffalo, State University of bd
North Carolina State University at Raleigh bdC
Northrop University bd
Notre Dame, University of bd
Oklahoma, University of bd
Parks College of St. Louis University bd
Pennsylvania State University bd
Polytechnic University bd
Princeton University bd
San Diego State University bd
Southern California, University of bdC
Syracuse University bd
Tennessee at Knoxville, University of bdC
Texas A&M University bdC
Texas at Arlington, University of bd
Texas at Austin, University of bdC
Tri-State University bdC
United States Naval Academy bd
Virginia, University of bd
West Virginia University bd

Aerospace Engineering and Mechanics

Minnesota, University of bd

Aerospace Engineering Sciences

Colorado at Boulder, University of bd

Aerospace Option in Mechanical Engineering[2]

Oklahoma State University bdC

Astronautical Engineering

United States Air Force Academy bd

AGRICULTURAL GROUP (ASAE)

(Programs in this group are accredited according to the program criteria for Agricultural and similarly named engineering programs.)

Agricultural and Irrigation Engineering

Utah State University bd

Agricultural Engineering

Arizona, University of bd
Arkansas State University bd
Arkansas, University of bd
Auburn University bdC
California Polytechnic State University, San Luis Obispo bdC
California State Polytechnic University, Pomona bd
California, Davis; University of bd
Clemson University bdC
Colorado State University bd
Cornell University bd
Florida, University of bdC
Georgia, University of bd
Idaho, University of bd

Illinois at Urbana-Champaign, University of bdC
Iowa State University bdC
Kansas State University bd
Kentucky, University of bd
Louisiana State University bd
Maryland, University of bdC
Michigan State University bd
Minnesota, University of bd
Mississippi State University bdC
Missouri-Columbia, University of bd
Montana State University bd
Nebraska-Lincoln, University of bd
New Mexico State University bd
North Dakota State University bd
Ohio State University bd
Oklahoma State University bdC
Oregon State University bd
Pennsylvania State University bd
Purdue University, West Lafayette bdC
Rutgers-The State University of New Jersey bd
South Dakota State University bd
Tennessee at Knoxville, University of bdC
Texas A&M University bdC
Texas Tech University bd
Virginia Polytechnic Institute and State University bdC
Washington State University bd
Wisconsin, Madison; University of bd
Wyoming, University of bd

Biological and Agricultural Engineering

North Carolina State University at Raleigh bdC

Biological Engineering

Mississippi State University bdC

Bio-Resource Engineering

Maine at Orono, University of bd

ARCHITECTURAL GROUP (ASCE Lead Society, with ASHRAE)

(Programs in this group are accredited according to the ABET general criteria and program criteria for nontraditional engineering programs.)

Architectural Engineering

California Polytechnic State University, San Luis Obispo bdC
Colorado at Boulder, University of bd
Kansas State University bd
Kansas, University of bd
Miami, University of bd
Milwaukee School of Engineering bd
North Carolina Agricultural & Technical State University bdC
Oklahoma State University bd
Pennsylvania State University bd
Tennessee State University bd
Texas at Austin, University of bdC
Wyoming, University of bd

BIOENGINEERING GROUP (IEEE Lead Society, with AIChE, ASAE, ASME, and NICE)

(Programs in this group are accredited according to the program criteria for Bio engineering and similarly named engineering programs.)

Bioengineering

Arizona State University bd
California, San Diego; University of bd
Illinois at Chicago, University of bd
Pennsylvania, University of bd
Texas A&M University bdC

[Biological and Agricultural Engineering]

[The program with this title is accredited according to the Agricultural Engineering program criteria only]

[Biological Engineering]

[The program with this title is accredited according to the Agricultural Engineering program criteria]

Biomedical Engineering

Boston University bd
Brown University bd
Case Western Reserve University bdC
Duke University bd
Iowa, University of bdC
Johns Hopkins University bd
Louisiana Tech University bd
Marquette University bdC
Northwestern University bdC
Rensselaer Polytechnic Institute bdC
Tulane University bd
Wright State University bde

Bio-Resource Engineering

[The program with this title is accredited according to the Agricultural Engineering program criteria.]

CERAMIC GROUP (NICE)

(Programs in this group are accredited according to the program criteria for Ceramic and similarly named engineering programs.)

Ceramic(s) Engineering

Alfred University, New York State College of Ceramics at bd
Clemson University bdC
Georgia Institute of Technology bdeC
Illinois at Urbana-Champaign, University of bdC
Iowa State University bd
Missouri-Rolla, University of bd
Ohio State University bd
Rutgers-The State University of New Jersey bd
Washington, University of bd

Ceramic Engineering Science

Alfred University, New York State College of Ceramics at bd

Ceramic Science and Engineering

Pennsylvania State University bd

Glass Engineering Science

Alfred University, New York State College of Ceramics at bd

CHEMICAL GROUP (AIChE)

(Programs in this group are accredited according to the program criteria for Chemical and similarly named engineering programs.)

Chemical Engineering

Akron, University of bdC
Alabama, University of bdC
Alabama in Huntsville, University of bde
Arizona State University bd
Arizona, University of bd
Arkansas, University of bd
Auburn University bdC
Brigham Young University bd
Brown University bd
Bucknell University bd
California Institute of Technology bd
California State Polytechnic University, Pomona bd
California State University, Long Beach bd
California, Berkeley; University of bd
California, Davis; University of bd
California, Los Angeles; University of bd
California, San Diego; University of bd
California, Santa Barbara; University of bd
Carnegie-Mellon University bd
Case Western Reserve University bdC
Christian Brothers College bd
Cincinnati, University of bC
Clarkson University bd
Clemson University bdC
Cleveland State University bdC
Colorado State University bd
Colorado at Boulder, University of bd
Columbia University bd
Connecticut, University of bd
Cooper Union, The bd
Cornell University bd
Dayton, University of bd
Delaware, University of bd
Detroit, University of bdC
Drexel University bdC
Florida A & M University/ Florida State University (FAMU/FSU) bd
Florida Institute of Technology bd
Florida, University of bdC
Georgia Institute of Technology bdC
Houston, University of bdC
Howard University bdC
Idaho, University of bd
Illinois Institute of Technology bdC
Illinois at Chicago, University of bd
Illinois at Urbana-Champaign, University of bdC
Iowa State University bdC
Iowa, University of bdC
Johns Hopkins University bd
Kansas State University bd
Kansas, University of bd
Kentucky, University of bd
Lafayette College bd
Lamar University bdC
Lehigh University bd
Louisiana State University bd
Louisiana Tech University bd
Lowell, University of bdC
Maine at Orono, University of bd
Manhattan College bd
Maryland, University of bdC
Massachusetts Institute of Technology bd

Massachusetts at Amherst, University of bd
Michigan State University bd
Michigan Technological University bd
Michigan, University of; Ann Arbor bd
Minnesota, University of bd
Mississippi State University bdC
Mississippi, University of bd
Missouri-Columbia, University of bd
Missouri-Rolla, University of bd
Montana State University bd
Nebraska-Lincoln, University of bd
Nevada-Reno, University of bd
New Hampshire, University of bd
New Jersey Institute of Technology bdeC
New Mexico State University bd
New Mexico, University of bd
New York at Buffalo, State University of bd
New York, City College of the City University of bd
North Carolina State University at Raleigh bdC
North Dakota, University of bd
Northeastern University bdC
Northwestern University bdC
Notre Dame, University of bd
Ohio State University bd
Ohio University bd
Oklahoma State University bdC
Oklahoma, University of bd
Oregon State University bd
Pennsylvania State University bd
Pennsylvania, University of bd
Pittsburgh, University of bd
Polytechnic University bd
Pratt Institute bdC
Princeton University bd
Puerto Rico, Mayaguez Campus; University of bdC
Purdue University, West Lafayette bdC
Rensselaer Polytechnic Institute bdC
Rhode Island, University of bd
Rice University bd
Rochester, University of bd
Rose-Hulman Institute of Technology bd
Rutgers-The State University of New Jersey bd
San Jose State University bd
South Alabama, University of bdC
South Carolina, University of bd
South Dakota School of Mines and Technology bd
South Florida, University of bd
Southern California, University of bdC
Southwestern Louisiana, University of bd
Stanford University bd
Stevens Institute of Technology bd
Syracuse University bd
Tennessee Technological University bdC
Tennessee at Knoxville, University of bdC
Texas A&I University bd
Texas A&M University bdC
Texas Tech University bd
Texas at Austin, University of bdC
Toledo, University of bd
Tufts University bd
Tulane University bd
Tulsa, University of bd
Tuskegee University bdC
Utah, University of bd
Vanderbilt University bd
Villanova University bd
Virginia Polytechnic Institute and State University bdC
Virginia, University of bd
Washington State University bd
Washington University bdC
Washington, University of bd
Wayne State University bdeC
West Virginia Institute of Technology bdC
West Virginia University bd
Widener University bdC
Wisconsin-Madison, University of bd
Worcester Polytechnic Institute bd
Wyoming, University of bd
Yale University bd
Youngstown State University bd

Chemical and Petroleum-Refining Engineering

Colorado School of Mines bd

CIVIL GROUP (ASCE)

(Programs in this group are accredited according to the program criteria for Civil and similarly named engineering programs.)

Civil and Environmental Engineering[3]

Vanderbilt University bd

Civil Engineering

Akron, University of bdC
Alabama, University of bdC
Alabama in Birmingham, University of bde
Alabama in Huntsville, University of bde
Alaska, Anchorage; University of bd
Alaska, Fairbanks; University of bd
Arizona State University bd
Arizona, University of bd
Arkansas, University of bd
Auburn University bdC
Bradley University bdC
Brigham Young University bd
Brown University bd
Bucknell University bd
California Polytechnic State University, San Luis Obispo bdC
California State Polytechnic University, Pomona bd
California State University, Chico bd
California State University, Fresno bd
California State University, Fullerton bde
California State University, Long Beach bd
California State University, Los Angeles bd
California State University, Sacramento bd
California, Berkeley; University of bd
California, Davis; University of bd
California, Irvine; University of bd
California, Los Angeles; University of bd
Carnegie-Mellon University bd
Case Western Reserve University bdC
Catholic University of America bd
Central Florida, University of bd
Christian Brothers College bd
Cincinnati, University of bC
Citadel, The bd
Clarkson University bd
Clemson University bdC
Cleveland State University bdC
Colorado State University bd
Colorado at Boulder, University of bd
Colorado at Denver, University of bd
Columbia University bd

[3]Joint with AAEE under Environmental Group

Connecticut, University of bd
Cooper Union, The bd
Cornell University bd
Dayton, University of bd
Delaware, University of bd
Detroit, University of bdC
District of Columbia, University of bd
Drexel University bdC
Duke University bd
Florida A&M University/Florida State University (FAMU/FSU) bd
Florida Institute of Technology bd
Florida International University bd
Florida, University of bdC
George Washington University bd
Georgia Institute of Technology bdeC
Gonzaga University bd
Hartford, University of bde
Hawaii at Manoa, University of bd
Houston, University of bdC
Howard University bdC
Idaho, University of bd
Illinois Institute of Technology bdC
Illinois at Chicago, University of bd
Illinois at Urbana-Champaign, University of bdC
Iowa State University bdC
Iowa, University of bdC
Johns Hopkins University bd
Kansas State University bd
Kansas, University of bd
Kentucky, University of bd
Lafayette College bd
Lamar University bdC
Lehigh University bd
Louisiana State University bd
Louisiana Tech University bd
Lowell, University of bdC
Loyola Marymount University bd
Maine at Orono, University of bd
Manhattan College bde
Marquette University bdeC
Maryland, University of bdC
Massachusetts Institute of Technology bdC
Massachusetts at Amherst, University of bd
Memphis State University bd
Merrimack College bdC
Miami, University of bd
Michigan State University bd
Michigan Technological University bd
Michigan, University of; Ann Arbor bd
Minnesota, University of bd
Mississippi State University bdC
Mississippi, University of bd
Missouri-Columbia, University of bd
Missouri-Columbia, University of (Kansas City) bd
Missouri-Rolla, University of bd
Montana State University bd
Nebraska-Lincoln, University of bd (This program is offered at both the Lincoln and Omaha campuses)
Nevada-Las Vegas, University of bd
Nevada-Reno, University of bd
New England College bd
New Hampshire, University of bd
New Haven, University of bde
New Jersey Institute of Technology bdeC
New Mexico State University bd
New Mexico, University of bd
New Orleans, University of bdC
New York at Buffalo, State University of bd
New York, City College of the City University of bde
North Carolina at Charlotte, University of bdC
North Carolina State University at Raleigh bdC
North Dakota State University bd
North Dakota, University of bd
Northeastern University bdeC

Northern Arizona University bdC
Northwestern University bdC
Norwich University bd
Notre Dame, University of bd
Ohio Northern University bd
Ohio State University bd
Ohio University bd
Oklahoma State University bdC
Oklahoma, University of bd
Old Dominion University bdC
Oregon State University bd
Pacific, University of the bdC
Pennsylvania State University bd
Pennsylvania, University of bd
Pittsburgh, University of bd
Polytechnic University bde
Portland State University bd
Portland, University of bd
Prairie View A&M University bd
Pratt Institute bdC
Princeton University bd
Puerto Rico, Mayaguez Campus; University of bdC
Purdue University, West Lafayette bdC
Rensselaer Polytechnic Institute bdC
Rhode Island, University of bd
Rice University bd
Rose-Hulman Institute of Technology bd
Rutgers-The State University of New Jersey bd
St. Martin's College bd
San Diego State University bd
San Francisco State University bd
San Jose State University bd
Santa Clara, University of bd
Seattle University bd
South Carolina, University of bd
South Dakota School of Mines and Technology bd
South Dakota State University bd
South Florida, University of bd
Southeastern Massachusetts University bd
Southern California, University of bdC
Southern Illinois University-Carbondale bd
Southern Illinois University-Edwardsville bd
Southern Methodist University bdC
Southern University and Agricultural and Mechanical College bdC
Southwestern Louisiana, University of bd
Stanford University bd
Stevens Institute of Technology bd
Syracuse University bd
Temple University bd
Tennessee State University bd
Tennessee Technological University bdC
Tennessee at Knoxville, University of bdC
Texas A&I University bd
Texas A&M University bdC
Texas Tech University bd
Texas at Arlington, University of bd
Texas at Austin, University of bdC
Texas at El Paso, University of bd
Texas at San Antonio, University of bd
Toledo, University of bd
Tri-State University bdC
Tufts University bd
Tulane University bd
Union College bde
United States Air Force Academy bd
United States Coast Guard Academy bd
United States International University bd
United States Military Academy bd
Utah State University bd
Utah, University of bd
Valparaiso University bd
Vermont, University of bd
Villanova University bd
Virginia Military Institute bd
Virginia Polytechnic Institute and State University bdC
Virginia, University of bd
Washington State University bd
Washington University bdeC
Washington, University of bd
Wayne State University bdeC
West Virginia Institute of Technology bdC
West Virginia University bd
Widener University bdC
Wisconsin-Madison, University of bd
Wisconsin-Milwaukee, University of bd
Wisconsin-Platteville, University of bd
Worcester Polytechnic Institute bd
Wyoming, University of bd
Youngstown State University bde

Construction Option in Civil Engineering

(See listing under Construction Group)

Structural Engineering

California, San Diego; University of bd

Surveying Engineering (Option in Civil and Environmental Engineering)[4]

Wisconsin-Madison, University of bd

COMPUTER GROUP (IEEE Lead Society, with AIChE and IIE)

(Programs in this group are accredited according to the program criteria for Computer and similarly named engineering programs.)

Computer and Electrical Engineering

Purdue University, West Lafayette bdC

Computer and Information Engineering Sciences

Florida, University of bdC

Computer and Systems Engineering

Rensselaer Polytechnic Institute bdC

Computer Engineering

Arizona, University of bd
Auburn University bdC
Boston University bd
Bridgeport, University of bdC
California State University, Chico bd
California State University, Sacramento bd
California, Santa Cruz; University of bd
Carnegie-Mellon University bd
Case Western Reserve University bdC
Central Florida, University of bd
Cincinnati, University of bC
Clemson University bdC
Florida Institute of Technology bd
George Washington University bd
Illinois at Urbana-Champaign, University of bdC
Iowa State University bdC

[4]Joint with AAEE under Environmental Group and ACSM under Surveying Group

Lehigh University bd
Louisiana State University bd
Miami, University of bd
Michigan, University of; Ann Arbor bd
Minnesota, Duluth; University of bd
Mississippi State University bdC
Missouri-Columbia, University of bd
New Mexico, University of bd
Oakland University bdC
Old Dominion University bdC
Oregon State University bd
Pacific, University of the bdC
Rochester Institute of Technology bC
Santa Clara University bd
South Florida, University of bd
Southeastern Massachusetts University bd
Southern Methodist University bdC
Stevens Institute of Technology bd
Syracuse University bd
Texas at Austin, University of bd
Washington, University of bd
Wright State University bde

Computer Engineering Option in Electrical Engineering

Tufts University bd

Computer Science

California, Berkeley; University of bd

Computer Science and Engineering

California State University, Long Beach bd
California, Davis; University of bd
California, Los Angeles; University of bd
Connecticut, University of bd
Illinois at Chicago, University of bd
Massachusetts Institute of Technology bdC
Milwaukee School of Engineering bd
Northern Arizona University bdC
Texas at Arlington, University of bd
Toledo, University of bd
Washington University bdC

Computer Systems Engineering

Arizona State University bd
Massachusetts at Amherst, University of bd
Western Michigan University, Kalamazoo Campus, bd

Electrical Engineering and Computer Science

(See listing under Electrical Group)

CONSTRUCTION GROUP (ASCE)

(Programs in this group are accredited according to the program criteria for Construction and similarly named engineering programs.)

Construction Engineering

Iowa State University bdC
Lawrence Institute of Technology bde
North Dakota State University bd

Construction Engineering and Management

Purdue University, West Lafayette bd

Construction Option in Civil Engineering

North Carolina State University at Raleigh bdC

ELECTRICAL AND ELECTRONIC(S) GROUP (IEEE)

(Programs in this group are accredited according to the program criteria for Electrical and similarly named engineering programs.)

Computer and Electrical Engineering

(See listing under Computer Group)

Computer Engineering Option in Electrical Engineering

(See listing under Computer Group)

Electric Power Engineering

Rensselaer Polytechnic Institute bdC

Electrical and Electronic(s) Engineering

California State University, Chico bd
California State University, Sacramento bdC
North Dakota State University bd
Northrop University bd
Oregon State University bd

Electrical Engineering

Akron, University of bdC
Alabama, University of bdC
Alabama in Birmingham, University of bde
Alabama in Huntsville, University of bde
Alaska, Fairbanks; University of bd
Arizona State University bd
Arizona, University of bd
Arkansas, University of bd
Auburn University bdC
Boston University bd
Bradley University bdC
Bridgeport, University of bdC
Brigham Young University bd
Brown University bd
Bucknell University bd
California Polytechnic State University, San Luis Obispo bdC
California State Polytechnic University, Pomona bd
California State University, Fresno bd
California State University, Fullerton bde
California State University, Long Beach bd
California State University, Los Angeles bd
California, Berkeley; University of bd
California, Davis; University of bd
California, Irvine; University of bd
California, Los Angeles; University of bd
California, San Diego; University of bd
California, Santa Barbara; University of bd
Carnegie-Mellon University bd
Case Western Reserve University bdC
Catholic University of America bd
Central Florida, University of bd
Christian Brothers College bd
Cincinnati, University of bC
Citadel, The bd
Clarkson University bd
Clemson University bdC
Cleveland State University bdC
Colorado at Boulder, University of bd
Colorado at Colorado Springs, University of bde
Colorado at Denver, University of bd
Colorado State University bd
Columbia University bd
Connecticut, University of bd
Cooper Union, The bd
Cornell University bd
Dayton, University of bd
Delaware, University of bd
Detroit, University of bC
District of Columbia, University of bd
Drexel University bdC
Duke University bd
Evansville, University of bdeC
Fairleigh Dickinson University, Teaneck Campus bde
Florida A&M University/Florida State University (FAMU/FSU) bd
Florida Atlantic University bd
Florida Institute of Technology bd
Florida International University bd
Florida, University of bdC
Gannon University bde
George Washington University bd
Georgia Institute of Technology bdeC
GMI Engineering and Management Institute bdC
Gonzaga University bd
Hartford, University of bde
Hawaii at Manoa, University of bd
Hofstra University bde
Houston, University of bdC
Howard University bdC
Idaho, University of bd
Illinois Institute of Technology bdC
Illinois at Chicago, University of bd
Illinois at Urbana-Champaign University of bdC
Indiana University-Purdue University at Indianapolis bdeC
Iowa State University bdC
Iowa, University of bdC
Johns Hopkins University bd
Kansas State University bd
Kansas, University of bd
Kentucky, University of bd
Lafayette College bde
Lamar University bdC
Lawrence Institute of Technology bdeC
Lehigh University bd
Louisiana State University bd
Louisiana Tech University bd
Lowell, University of bdC
Loyola Marymount University bd
Maine at Orono, University of bd
Manhattan College bde
Mankato State University bd
Marquette University bdeC
Maryland, University of bdC
Massachusetts at Amherst, University of bd
Memphis State University bd
Merrimack College bdC
Miami, University of bd
Michigan State University bd
Michigan Technological University bd
Michigan, University of; Ann Arbor bd
Michigan-Dearborn, University of bdC
Milwaukee School of Engineering bd
Minnesota, University of bd
Mississippi State University bdC
Mississippi, University of bd
Missouri-Columbia, University of bd
Missouri-Columbia, University of (Kansas City) bd
Missouri-Rolla, University of bd
Montana State University bd
Nebraska-Lincoln, University of bd
Nevada-Las Vegas, University of bd
Nevada, Reno; University of bd
New Hampshire, University of bd
New Haven, University of bde
New Jersey Institute of Technology bdeC
New Mexico State University bd
New Mexico, University of bd
New Orleans, University of bdC
New York at Binghamton, State University of bd
New York at Buffalo, State University of bd
New York at New Paltz, State University of bde
New York at Stony Brook, State University of bd
New York Institute of Technology (Metropolitan Center) bde
New York Institute of Technology (Old Westbury) bde
New York Maritime College, State University of bd
New York, City College of the City University of bde
North Carolina Agricultural and Technical State University bdC
North Carolina at Charlotte, University of bdC
North Carolina State University at Raleigh bdC
North Dakota, University of bd
Northeastern University bdeC
Northern Arizona University bdC
Northwestern University bdC
Norwich University bd
Notre Dame, University of bd
Oakland University bdC
Ohio Northern University bd
Ohio State University bd
Ohio University bd
Oklahoma State University bdC
Oklahoma, University of bd
Old Dominion University bdC
Pacific, University of the bdC
Pennsylvania State University bdC
Pennsylvania, University of bd
Pittsburgh, University of bde
Polytechnic University bde
Portland State University bd
Portland, University of bd
Prairie View A&M University bd
Pratt Institute bdC
Princeton University bd
Puerto Rico, Mayaguez Campus; University of bdC
Purdue University, West Lafayette bdC
Purdue University Calumet bdeC
Rensselaer Polytechnic Institute bdC
Rhode Island, University of bd
Rice University bd
Rochester Institute of Technology bC
Rochester, University of bd
Rose-Hulman Institute of Technology bd
Rutgers-The State University of New Jersey bd
St. Cloud State University bd
St. Mary's University bd
San Diego State University bd
San Francisco State University bd
San Jose State University bd
Santa Clara University bd
Seattle University bd
Seattle Pacific University bd
South Alabama, University of bdC
South Carolina, University of bd
South Dakota School of Mines and Technology bd
South Dakota State University bd
South Florida, University of bd
Southeastern Massachusetts University bd
Southern California, University of bdC
Southern Illinois University at Carbondale bd
Southern Illinois University at Edwardsville bd
Southern Methodist University bdC

Southern University and Agricultural and Mechanical College bdC
Southwestern Louisiana, University of bd
Stanford University bd
Stevens Institute of Technology bd
Syracuse University bd
Temple University bd
Tennessee State University bd
Tennessee Technological University bdC
Tennessee at Knoxville, University of bdC
Texas A&I University bd
Texas A&M University bdC
Texas Tech University bd
Texas at Arlington, University of bd
Texas at Austin, University of bdC
Texas at El Paso, University of bd
Texas at San Antonio, University of bd
Toledo, University of bd
Tri-State University bdC
Tufts University bd
Tulane University bd
Tulsa, University of bd
Tuskegee University bdC
Union College bde
United States Air Force Academy bd
United States Coast Guard Academy bd
United States Military Academy bd
United States Naval Academy bd
Utah State University bd
Utah, University of bd
Valparaiso University bd
Vanderbilt University bd
Vermont, University of bd
Villanova University bde
Virginia Military Institute bd
Virginia Polytechnic Institute and State University bdC
Virginia, University of bd
Washington State University (Pullman) bd
Washington State University (Richland) be
Washington University bdC
Washington, University of bd
Wayne State University bdeC
West Coast University be*
West Virginia Institute of Technology bdC
West Virginia University bd
Western Michigan University, Kalamazoo Campus bd
Western New England College bd
Wichita State University bde
Widener University bdC
Wilkes College bde
Wisconsin-Madison, University of bd
Wisconsin-Milwaukee, University of bd
Wisconsin-Platteville, University of bd
Worcester Polytechnic Institute bd
Wright State University bde
Wyoming, University of bd
Yale University bd
Youngstown State University bd

Electrical Engineering and Computer Science

Colorado at Boulder, University of bd

Electrical Science and Engineering

Massachusetts Institute of Technology bdC

Electronic(s) Engineering

California Polytechnic State University, San Luis Obispo bdC

***See note under Part I regarding period of accreditation.**

George Mason University bd
Monmouth College bde

Microelectronic Engineering

Rochester Institute of Technology bC

ENGINEERING (GENERAL) GROUP (EAC of ABET, with ASCE, ASHRAE, ASME, IIE, and SAE)

(Programs in this group are accredited according to the ABET general criteria and program criteria for nontraditional engineering programs.)

Engineering

Arkansas State University bd
Arkansas Tech University bd
Baylor University bd
California State University, Northridge bde
Calvin College bd
Colorado School of Mines bd
Dartmouth College bd
Harvey Mudd College bd
LeTourneau College bd
Maryland, University of bdC
McNeese State University bd
Michigan Technological University bd
Oklahoma, University of bd
Stevens Institute of Technology bd
Swarthmore College bd
Tennessee at Chattanooga, University of bdeC
Walla Walla College bd
Widener University bdC

Engineering and Public Policy

Carnegie-Mellon University bd
Washington University bdC

Engineering Interdisciplinary Studies

Arizona State University bd

Engineering Special Studies

Arizona State University bd

General Engineering

Idaho State University bd
Illinois at Urbana-Champaign, University of bdC
Oklahoma State University bdC

ENGINEERING MANAGEMENT GROUP (IIE Lead Society, with AIChE, ASCE, ASME, SME and SPE)

(Programs in this group are accredited according to the ABET general criteria and program criteria for nontraditional engineering programs.)

Engineering Management

Missouri-Rolla, University of bd
United States Military Academy bd

Industrial and Management Engineering

(See listing under Industrial Group)

Industrial Engineering and Management

(See listing under Industrial Group)

ENGINEERING MECHANICS GROUP (ASME Lead Society, with ASCE and SAE)

(Programs in this group are accredited according to the program criteria for Engineering Mechanics and similarly named engineering programs.)

Engineering Mechanics

Cincinnati, University of bC
Columbia University bd
Illinois at Urbana-Champaign, University of bdC
Johns Hopkins University bd
Lehigh University bd
Missouri-Rolla, University of bd
United States Air Force Academy bd
Wisconsin-Madison, University of bd

Engineering Science and Mechanics

Georgia Institute of Technology bdeC
Virginia Polytechnic Institute and State University bdC

Mechanical Engineering and Applied Mechanics

(See listing under Mechanical Group)

ENGINEERING PHYSICS/ ENGINEERING SCIENCE GROUP (EAC of ABET, with AIAA, ASCE, ASME, and IEEE)

(Programs in this group are accredited according to the ABET general criteria and program criteria for nontraditional engineering programs.)

Engineering and Applied Science

California Institute of Technology bd

Engineering Physics

Colorado School of Mines bd
Cornell University bd
Kansas, University of bd
Maine at Orono, University of bd
Oklahoma, University of bd
Pacific, University of the bdC
Princeton University bd
Stevens Institute of Technology bd
Texas Tech University bd
Toledo, University of bd
Tulsa, University of bd
Wright State University bde

Engineering Science(s)

Colorado State University bd
Florida, University of bdC
Harvard University bd
Hofstra University bde
Iowa State University bdC
Michigan-Ann Arbor, University of bd
Montana College of Mineral Science and Technology bd
New York at Stony Brook, State University of bd
Pennsylvania State University bd
Staten Island, College of; CUNY bde
Tennessee at Knoxville, University of bdC

Trinity University bd
United States Air Force Academy bd

Engineering Science and Mechanics

(See listing under Engineering Mechanics Group)

Fluid & Thermal Engineering Sciences

Case Western Reserve University bdC

ENVIRONMENTAL AND SANITARY GROUP (AAEE Lead Society, with AIChE, ASCE, ASHRAE, ASME, and SAE)

(Programs in this group are accredited according to the program criteria for Environmental, Sanitary, and similarly named engineering programs.)

Civil and Environmental Engineering[a]

Vanderbilt University bd

Environmental Engineering

California Polytechnic State University, San Luis Obispo bdC
Central Florida, University of bd
Florida Institute of Technology bd
Florida, University of bdC
Michigan Technological University bd
Montana College of Mineral Science and Technology bd
Northwestern University bdC
Pennsylvania State University bdC
Rensselaer Polytechnic Institute bdC

Environmental Resources Engineering

Humboldt State University bd

Surveying Engineering (Option in Civil and Environmental Engineering)[b]

Wisconsin-Madison, University of bd

FOREST GROUP (ASAE Lead Society, with AIChE)

(Programs in this group are accredited according to the program criteria for Agricultural and similarly named engineering programs.)

Forest Engineering

Maine at Orono, University of bd
New York College of Environmental Science and Forestry; State University of bd

[a]Joint with ASCE under Civil Group
[b]Joint with ASCE under Civil Group and ACSM under Surveying Group
*See note under Part I regarding period of accreditation.

GEOLOGICAL AND GEOPHYSICAL GROUP (SME-AIME)

(Programs in this group are accredited according to the program criteria for Geological and similarly named engineering programs.)

Geological Engineering

Alaska, Fairbanks; University of bd
Arizona, University of bd
Colorado School of Mines bd
Idaho, University of bd
Michigan Technological University bd
Minnesota, University of bd
Mississippi, University of bd
Missouri-Rolla, University of bd
Montana College of Mineral Science and Technology bd
Nevada, Reno; University of bd
New Mexico Institute of Mining and Technology bd
New Mexico State University bd
North Dakota, University of bd
Princeton University bd
South Dakota School of Mines and Technology bd
Utah, University of bd
Washington State University bd

Geophysical Engineering

Colorado School of Mines bd
Montana College of Mineral Science and Technology bd

INDUSTRIAL GROUP (IIE)

(Programs in this group are accredited according to the program criteria for Industrial and similarly named engineering programs.)

Industrial and Management Engineering

Montana State University bd
Rensselaer Polytechnic Institute bdC

Industrial and Operations Engineering

Michigan, University of; Ann Arbor bd

Industrial and Systems Engineering

Alabama in Huntsville, University of bde
Florida, University of bdC
Michigan-Dearborn, University of bdC
Ohio State University bd
Ohio University bd
San Jose State University bd
Southern California, University of bdC

Industrial Engineering

Alabama, University of bdC
Alfred University, New York State College of Ceramics at bd
Arizona State University bd
Arizona, University of bd
Arkansas, University bd
Auburn University bdC
Bradley University bdC
California Polytechnic State University, San Luis Obispo bdC
California State Polytechnic University, Pomona bd
California State University, Fresno bd
California, Berkeley; University of bd
Central Florida, University of bd
Cincinnati, University of bC
Clemson University bdC
Cleveland State University bdC
Columbia University bd
Fairleigh Dickinson University, Teaneck Campus bde*
Florida International University bd
Georgia Institute of Technology bdeC
GMI Engineering and Management Institute bdC
Houston, University of bdC
Illinois at Chicago, University of bd
Illinois at Urbana-Champaign, University of bdC
Iowa State University bdC
Iowa, University of bdC
Kansas State University bd
Lamar University bdC
Lehigh University bd
Louisiana State University bd
Louisiana Tech University bd
Marquette University bdeC
Miami, University of bd
Mikwaukee School of Engineering bd
Minnesota, Duluth; University of bd
Mississippi State University bdC
Missouri-Columbia, University of bd
Nebraska-Lincoln, University of bd
New Haven, University of bde
New Jersey Institute of Technology bdeC
New Mexico State University bd
New York at Buffalo, State University of bd
North Carolina Agricultural and Technical State University bdC
North Carolina State University at Raleigh bdC
North Dakota State University bd
Northeastern University bdC
Northwestern University bdC
Oklahoma, University of bd
Oregon State University bd
Pennsylvania State University bdC
Pittsburgh, University of bd
Polytechnic University bde
Puerto Rico, Mayaguez Campus; University of bdC
Purdue University, West Lafayette bdC
Rhode Island, University of bd
Rochester Institute of Technology bC
Rutgers-The State University of New Jersey bd
St. Mary's University bd
South Florida, University of bd
Stanford University bd
Tennessee Technological University bdC
Tennessee at Knoxville, University of bdC
Texas A&M University bdC
Texas Tech University bd
Texas at Arlington, University of bd
Texas at El Paso, University of bd
Toledo, University of bd
Washington, University of bd
Wayne State University bdeC
West Virginia University bd
Western Michigan University, Kalamazoo Campus bd
Western New England College bd
Wichita State University bde
Wisconsin-Madison, University of bd
Wisconsin-Milwaukee, University of bd
Wisconsin-Platteville, University of bd
Youngstown State University bd

Industrial Engineering and Management

Oklahoma State University bdC

Industrial Engineering and Operations Research

Cornell University bd
Massachusetts at Amherst, University of bd
Virginia Polytechnic Institute and State University of bdC

Manufacturing Engineering Option in Industrial Engineering[7]

Kansas State University bd
Oregon State University bd

MANUFACTURING GROUP (SME)

(Programs in this group are accredited according to the program criteria for Manufacturing and similarly named engineering programs.)

Manufacturing Engineering

Boston University bd
Bradley University bdC
California State Polytechnic University, Pomona bd
Utah State University bd

Manufacturing Engineering Option in Industrial Engineering[8]

Kansas State University bd
Oregon State University bd

MATERIALS GROUP (TMS Lead Society, with AIChE, ASME, and NICE)

(Programs in this group are accredited according to the program criteria for Metallurgical, Materials, and similarly named engineering programs.)

Materials and Metallurgical Engineering

(See listing under Metallurgical Group)

Materials Engineering

Alabama in Birmingham, University of bde
Auburn University bdC
Brown University bd
California, Los Angeles; University of bd
Drexel University bdC
North Carolina State University at Raleigh bdC
Rensselaer Polytechnic Institute bdC
San Jose State University bd
Virginia Polytechnic Institute and State University bdC
Wilkes College bde
Wisconsin-Milwaukee, University of bd

Materials Science and Engineering

Arizona, University of bd
Case Western Reserve University bdC
Cornell University bd
Florida, University of bdC
Johns Hopkins University bd
Lehigh University bd
Massachusetts Institute of Technology bdC
Michigan State University bd
Michigan, University of; Ann Arbor bd
Minnesota, University of bd
Northwestern University bdC
Pennsylvania, University of bd
Pittsburgh, University of bd
Rice University bd
Utah, University of bd
Washington State University bd
Wright State University bde

Materials Science and Engineering Option in Metallurgical Engineering

(See listing under Metallurgical Group)

Metallurgical Engineering and Materials Science

(See listing under Metallurgical Group)

Metals Science and Engineering

(See listing under Metallurgical Group)

MECHANICAL GROUP (ASME)

(Programs in this group are accredited according to the program criteria for Mechanical and similarly named engineering programs.)

Aerospace Option in Mechanical Engineering[9]

Oklahoma State University bdC

Mechanical Engineering

Akron, University of bdC
Alabama, University of bdC
Alabama in Birmingham, University of bde
Alabama in Huntsville, University of bde
Alaska, Fairbanks; University of bd
Arizona State University bd
Arizona, University of bd
Arkansas, University of bd
Auburn University bdC
Boston University bd
Bradley University bdC
Bridgeport, University of bdC
Brigham Young University bd
Brown University bd
Bucknell University bd
California Polytechnic State University, San Luis Obispo bdC
California State Polytechnic University, Pomona bd
California State University, Chico bd
California State University, Fresno bd
California State University, Fullerton bde
California State University, Long Beach bd
California State University, Los Angeles bd
California State University, Sacramento bd
California, Berkeley; University of bd
California, Davis; University of bd
California, Irvine; University of bd
California, Los Angeles; University of bd
California, San Diego; University of bd
California, Santa Barbara; University of bd
Carnegie-Mellon University bd
Case Western Reserve University bdC
Catholic University of America bd
Central Florida, University of bd
Christian Brothers College bd
Cincinnati, University of bC
Clarkson University bd
Clemson University bdC
Cleveland State University bdC
Colorado State University bd
Colorado at Boulder, University of bd
Colorado at Denver, University of bd
Columbia University bd
Connecticut, University of bd
Cooper Union, The bd
Cornell University bd
Dayton, University of bd
Delaware, University of bd
Detroit, University of bdC
District of Columbia, University of bd
Drexel University bdC
Duke University bd
Evansville, University of bdeC
Fairleigh Dickinson University, Teaneck Campus bde*
Florida A&M University/Florida State University (FAMU/FSU) bd
Florida Atlantic University bd
Florida Institute of Technology bd
Florida International University bd
Florida, University of bdC
Gannon University bde
George Washington University bd
Georgia Institute of Technology bdeC
GMI Engineering and Management Institute bdC
Gonzaga University bd
Hartford, University of bde
Hawaii at Manoa, University of bd
Hofstra University bde
Houston, University of bdC
Howard University bdC
Idaho, University of bd
Illinois Institute of Technology bdC
Illinois at Chicago, University of bd
Illinois at Urbana-Champaign, University of bdC
Indiana University-Purdue University at Indianapolis bdeC
Iowa State University bdC
Iowa, University of bdC
Kansas State University bd
Kansas, University of bd
Kentucky, University of bd
Lafayette College bde
Lamar University bdC
Lawrence Institute of Technology bdeC
Lehigh University bd
Louisiana State University bd
Louisiana Tech University bd
Lowell, University of bdC
Loyola Marymount University bd
Maine at Orono, University of bd
Manhattan College bde
Marquette University bdeC
Maryland, University of bdC
Massachusetts Institute of Technology bdC
Massachusetts at Amherst, University of bd
Memphis State University bd
Miami, University of bd
Michigan State University bd
Michigan Technological University bd
Michigan, University of; Ann Arbor bd
Michigan-Dearborn, University of bdC
Milwaukee School of Engineering bd
Minnesota, University of bd
Mississippi State University bdC
Mississippi, University of bd
Missouri-Columbia, University of bd

[7]Joint with SME under Manufacturing Group
[8]Joint with IIE under Industrial Group
[9]Joint with AIAA under Aerospace Group
*See note under Part I regarding period of accreditation.

Missouri-Columbia, University of (Kansas City) bd
Missouri-Rolla, University of bd
Montana State University bd
Nebraska-Lincoln, University of bd
Nevada-Las Vegas, University of bd
Nevada-Reno; University of bd
New Hampshire, University of bd
New Haven, University of bde
New Jersey Institute of Technology bdeC
New Mexico State University bd
New Mexico, University of bd
New Orleans, University of bdC
New York at Binghamton, State University of bd
New York at Buffalo, State University of bd
New York at Stony Brook, State University of bd
New York Institute of Technology (Old Westbury) bd
New York, City College of the City University of bde
North Carolina Agricultural & Technical State University bdC
North Carolina at Charlotte, University of bdC
North Carolina State University at Raleigh bdC
North Dakota State University bd
North Dakota, University of bd
Northeastern University bdeC
Northern Arizona University bdC
Northrop University bd*
Northwestern University bdC
Norwich University bd
Notre Dame, University of bd
Oakland University bdC
Ohio Northern University bd
Ohio State University bd
Ohio University bd
Oklahoma State University bdC
Oklahoma, University of bd
Old Dominion University bdC
Oregon State University bd
Pacific, University of the bdC
Pennsylvania State University bdC
Pittsburgh, University of bde
Polytechnic University bde
Portland State University bd
Portland, University of bd
Prairie View A&M University bd
Pratt Institute bdC
Princeton University bd
Puerto Rico, Mayaguez Campus; University of bdC
Purdue University, West Lafayette bdC
Purdue University Calumet bdeC
Rensselaer Polytechnic Institute bdC
Rhode Island, University of bd
Rice University bd
Rochester Institute of Technology bC
Rochester, University of bd
Rose-Hulman Institute of Technology bd
Rutgers-The State University of New Jersey bd
San Diego State University bd
San Francisco State University bd
San Jose State University bd
Santa Clara University bd
Seattle University bd
South Alabama, University of bdC
South Carolina, University of bd
South Dakota School of Mines & Technology bd
South Dakota State University bd
South Florida, University of bd
Southeastern Massachusetts University bd
Southern California, University of bdC

*See note under Part I regarding period of accreditation.

Southern Illinois University-Carbondale bd
Southern Methodist University bdC
Southern University and Agricultural and Mechanical College bdC
Southwestern Louisiana, University of bd
Stanford University bd
Stevens Institute of Technology bd
Syracuse University bd
Temple University bd
Tennessee State University bd
Tennessee Technological University bdC
Tennessee at Knoxville, University of bdC
Texas A&I University bd
Texas A&M University bdC
Texas Tech University bd
Texas at Arlington, University of bd
Texas at Austin, University of bdC
Texas at El Paso, University of bd
Texas at San Antonio, University of bd
Toledo, University of bd
Tri-State University bdC
Tufts University bd
Tulane University bd
Tulsa, University of bd
Tuskegee University bdC
Union College bde
United States Military Academy bd
United States Naval Academy bd
Utah State University bd
Utah, University of bd
Valparaiso University bd
Vanderbilt University bd
Vermont, University of bd
Villanova University bde
Virginia Military Institute bd
Virginia Polytechnic Institute and State University bdC
Virginia, University of bd
Washington State University (Pullman) bd
Washington State University (Richland) be
Washington University bdeC
Washington, University of bd
Wayne State University bdeC
West Coast University be*
West Virginia Institute of Technology bdC
West Virginia University bd
Western Michigan University, Kalamazoo Campus, bd
Western New England College bd
Wichita State University bde
Widener University bdC
Wisconsin-Madison, University of bd
Wisconsin-Milwaukee, University of bd
Wisconsin-Platteville, University of bd
Worcester Polytechnic Institute bd
Wright State University bde
Wyoming, University of bd
Yale University bd
Youngstown State University bde

Mechanical Engineering and Applied Mechanics

Pennsylvania, University of bd

METALLURGICAL GROUP (TMS Lead Society, with SME-AIME)

(Programs in this group are accredited according to the program criteria for Metallurgical, Materials, and similarly named engineering programs.)

Extractive Metallurgical Engineering

Minnesota, University of bd

Materials and Metallurgical Engineering

Stevens Institute of Technology bd

Materials Science and Engineering Option in Metallurgical Engineering

Michigan Technological University bd

Metallurgical Engineering

Alabama, University of bdC
California Polytechnic State University, San Luis Obispo bdC
Cincinnati, University of bC
Colorado School of Mines bd
Columbia University bd
Idaho, University of bd
Illinois Institute of Technology bdC
Illinois at Chicago, University of bd
Illinois at Urbana-Champaign, University of bdC
Iowa State University bdC
Kentucky, University of bd
Lafayette College bd*
Missouri-Rolla, University of bd
Montana College of Mineral Science and Technology bd
Nevada, Reno; University of bd
New Mexico Institute of Mining and Technology bd*
Notre Dame, University of bd
Ohio State University bd
Pittsburgh, University of bd
Polytechnic University bd
Purdue University bdC
South Dakota School of Mines and Technology bd
Tennessee at Knoxville, University of bdC
Texas at El Paso, University of bd
Utah, University of bd
Washington, University of bd
Wayne State University bdeC
Wisconsin-Madison, University of bd

Metallurgical Engineering and Materials Science

Carnegie-Mellon University bd

Metals Science and Engineering

Pennsylvania State University bd

MINERAL GROUP (SME-AIME Lead Society, with TMS)

(Programs in this group are accredited according to the program criteria for either Metallurgical or Mining and similarly named engineering programs, and/or ABET general criteria.)

Mineral Engineering

Alabama, University of bdC
California, Berkeley; University of bd

Mineral Processing Engineering

Michigan Technological University bd (Option in Metallurgical Eng.)
Montana College of Mineral Science and Technology bd
Pennsylvania State University bdC (Option in Mining Engineering)

MINING GROUP (SME-AIME)

(Programs in this group are accredited according to the program criteria for Mining and similarly named engineering programs.)

Mining Engineering

Alaska, Fairbanks; University of bd
Arizona, University of bd
Colorado School of Mines bd
Columbia University bd
Idaho, University of bd
Kentucky, University of bd
Michigan Technological University bd
Missouri-Rolla, University of bd
Montana College of Mineral Science and Technology bd
Nevada, Reno; University of bd
New Mexico Institute of Mining and Technology bd
Pennsylvania State University (Mining Option) bdC
South Dakota School of Mines and Technology bd
Southern Illinois University-Carbondale bd
Utah, University of bd
Virginia Polytechnic Institute and State University bdC
West Virginia University bd

NAVAL ARCHITECTURE AND MARINE GROUP (SNAME)

(Programs in this group are accredited according to the program criteria for Naval Architecture, Marine, and similarly named engineering programs.)

Marine Engineering

New York Maritime College, State University of bd
Texas A&M University at Galveston bd
United States Coast Guard Academy bd
United States Naval Academy bd

Marine Engineering Systems

United States Merchant Marine Academy bC

Naval Architecture

California, Berkeley; University of bd
New York Maritime College, State University of bd
United States Naval Academy bd

Naval Architecture and Marine Engineering

Massachusetts Institute of Technology bdC
Michigan, University of; Ann Arbor bd
New Orleans, University of bdC
Webb Institute of Naval Architecture bd

[10]Joint with AIAA under Aerospace Group
[11]Joint with ASCE under Civil Group and AAEE under Environmental Group
*See note under Part I regarding period of accreditation.

NUCLEAR GROUP (ANS)

(Programs in this group are accredited according to the program criteria for Nuclear and similarly named engineering programs.)

Nuclear Engineering

Arizona, University of bd
California, Berkeley; University of bd
California, Santa Barbara; University of bd
Florida, University of bdC
Georgia Institute of Technology bdeC
Illinois at Urbana-Champaign, University of bdC
Iowa State University bdC
Kansas State University bd
Lowell, University of bdC
Maryland, University of bdC
Massachusetts Institute of Technology bdC
Michigan, University of; Ann Arbor bd
Mississippi State University bdC
Missouri-Rolla, University of bd
New Mexico, University of bd
North Carolina State University at Raleigh bdC
Oregon State University bd
Pennsylvania State University bd
Purdue University, West Lafayette bdC
Rensselaer Polytechnic Institute bdC
Tennessee at Knoxville, University of bdC
Texas A&M University bdC
Virginia, University of bd
Wisconsin-Madison, University of bd

Nuclear and Power Engineering

Cincinnati, University of bC

OCEAN GROUP (SNAME Lead Society, with ASCE and IEEE)

(Programs in this group are accredited according to the program criteria for Ocean and similarly named engineering programs.)

Aerospace and Ocean Engineering[10]

Virginia Polytechnic Institute and State University bdC

Ocean Engineering

Florida Atlantic University bd
Florida Institute of Technology bd
Massachusetts Institute of Technology bdC
Texas A&M University bdC
United States Naval Academy bd

PETROLEUM GROUP (SPE)

(Programs in this group are accredited according to the program criteria for Petroleum and similarly named engineering programs.)

[Chemical and Petroleum-Refining Engineering]

[The program with this title is accredited according to the Chemical Engineering program criteria only]

Natural Gas Engineering

Texas A&I University bd

Petroleum Engineering

Colorado School of Mines bd
Kansas, University of bd
Louisiana State University bd
Louisiana Tech University bd
Marietta College bd
Mississippi State University bdC
Missouri-Rolla, University of bd
Montana College of Mineral Science and Technology bd
New Mexico Institute of Mining and Technology bd*
Oklahoma, University of bd
Southern California, University of bdC
Southwestern Louisiana, University of bd
Stanford University bd
Texas A&M University bdC
Texas Tech University bd
Texas at Austin, University of bdC
Tulsa, University of bd
Wyoming, University of bd

Petroleum and Natural Gas Engineering

Pennsylvania State University bd
West Virginia University bd

PLASTICS ENGINEERING GROUP (EAC of ABET, with AIChE, SAE and SME)

Lowell, University of bdC

SURVEYING ENGINEERING GROUP (ACSM Lead Society, with ASCE)

(Programs in this group are accredited according to the program criteria for Surveying and similarly named engineering programs.)

Surveying Engineering

California State University, Fresno bd
Iowa State University bdC*
Maine at Orono, University of bd
Purdue University, West Lafayette bd

Surveying Engineering Option in Civil and Environmental Engineering[11]

Wisconsin-Madison, University of bd

SYSTEMS GROUP (EAC of ABET, with IIE, IEEE, SAE, and SME)

(Programs in this group are accredited according to the ABET general criteria and program criteria for nontraditional engineering programs.)

Computer & Systems Engineering

(See listing under Computer Group)

Industrial and Systems Engineering

(See listing under Industrial Group)

Systems Analysis and Engineering

George Washington University bd

Systems and Control Engineering

California, San Diego; University of bd

Case Western Reserve University
bdC

Systems Engineering

Arizona, University of bd
Boston University bd
Oakland University bdC
United States Naval Academy bd
Virginia, University of bd

System(s) Science and Engineering

Pennsylvania, University of bd
Washington University bdC

WELDING GROUP (TMS Lead Society, with ASME and SME)

(Programs in this group are accredited according to the ABET general criteria and program criteria for nontraditional engineering programs.)

Welding Engineering

Ohio State University bd

OTHERS (EAC of ABET)

(Programs in this group are accredited according to the ABET general criteria and program criteria for nontraditional engineering programs.)

Fire Protection Engineering

Maryland, University of bdC

Food Process Engineering

Purdue University, West Lafayette
bdC

Polymer Science and Engineering

Case Western Reserve University
bdC

Radiological Health Engineering

Texas A&M University bdC

Textile Engineering

Georgia Institute of Technology
bdeC

APPENDIX B

Accredited Engineering Technology Programs in the United States

Part II-A
Accredited Programs Leading to Associate Degrees in Engineering Technology, 1989
By Program Area

The listings below are grouped according to the title of the program as reported by the institution and accredited by the Technology Accreditation Commission (TAC) of the Accreditation Board for Engineering and Technology (ABET). Options under overall program titles are listed with the major programs.

The letters in parentheses after the program titles indicate the Participating Bodies of ABET that have been assigned curricular responsibility or designated as lead society for assisting ABET in the preparation of program criteria and the appointment of program evaluators for programs within their areas of professional disciplinary competence, as noted at the end of the Bachelor's Degree listing. Lead societies have primary but not exclusive responsibility for assigned program areas. Where no society has been designated, ABET itself has curricular responsibility.

This list includes ONLY ASSOCIATE DEGREE PROGRAMS. For bachelor's degree programs, refer to the separate listing that follows. Users should note that similar or related programs may be listed under variant titles. The entire list should be scanned for such programs.

Characteristics of the accredited programs are indicated after each program listing by combinations of the following symbols:

a—Associate program d—Day program e—Evening program de—Day and evening program

AEROSPACE GROUP (AIAA)

(Programs in this group are accredited according to the ABET general criteria for engineering technology programs.)

Aeronautical Engineering Technology

College of Aeronautics (Design Option) ade
College of Aeronautics (Maintenance Option) ade

Aeronautical Technology

Wentworth Institute of Technology ad

Electronics Option in Aeronautical Engineering Technology[1]

(See listing under Electrical and Electronic Group)

AIR CONDITIONING GROUP (ASHRAE)

(Programs in this group are accredited according to the ABET general criteria for engineering technology programs.)

Air Conditioning Engineering Technology

New York, State University of, College of Technology, Alfred ad
New York, State University of, College of Technology, Canton ad
New York, State University of, College of Technology, Farmingdale ad

[1]Joint with IEEE under Electrical and Electronic Group

ARCHITECTURAL GROUP (ASCE Lead Society, with ASHRAE)

(Programs in this group are accredited according to the ABET general criteria for engineering technology programs.)

Architectural and Building Engineering Technology

(See listing under Civil and Construction Group)

Architectural and Building Technology

(See listing under Civil and Construction Group)

Architectural and Civil Engineering Technology

(See listing under Civil and Construction Group)

Architectural Engineering Technology

Bluefield State College ad
District of Columbia, University of the; Van Ness Campus ade
Franklin Institute of Boston ad
Greenville Technical College ade
John Tyler Community College ad
Memphis, State Technical Institute at ade
Midlands Technical College ad
Nashville State Technical Institute ade
New Hampshire Technical Institute ade
New York, State University of, College of Technology, Alfred ad
Norwalk State Technical College ade
Pennsylvania State University, Fayette Campus ad
Pennsylvania State University, Worthington-Scranton Campus ad
Southern College of Technology ade
Wentworth Institute of Technology ad

Architectural Technology

Central Piedmont Community College ade
Cincinnati, University of; OMI College of Applied Science ade
Indiana University-Purdue University at Fort Wayne ade
Purdue University Calumet ade
Rochester Institute of Technology ad
Wentworth Institute of Technology ae

Architectural Drafting and Design Technology

Catawba Valley Technical College ad

Construction-Architectural Engineering Technology

(See listing under Civil and Construction Group)

AUTOMOTIVE GROUP (SAE)

(Programs in this group are accredited according to the ABET general criteria for engineering technology programs.)

Automotive Engineering Technology

New York, State University of, College of Technology, Farmingdale ad

Reprinted by permission of the Accreditation Board for Engineering and Technology, Inc.

BIOENGINEERING TECHNOLOGY GROUP (IEEE Lead Society, with AIChE, ASAE, ASME & NICE)

(Programs in this group are accredited according to the ABET program criteria for Bioengineering Technology and similarly named programs.)

Biomedical Engineering Technology

Cincinnati Technical College ade
Colorado Technical College, Colorado Springs Campus ade
Memphis, State Technical Institute at ade
New York, State University of, College of Technology, Farmingdale, ad

Biomedical Equipment Engineering Technology

Pennsylvania State University, New Kensington Campus ad
Pennsylvania State University, Wilkes-Barre Campus ad

Biomedical Equipment Option of Electrical/Electronics Engineering Technology

Owens Technical College ade

CERAMIC GROUP (NICE)

(Programs in this group are accredited according to the ABET general criteria for engineering technology programs).

Ceramic Engineering Technology

Hocking Technical College ad

CHEMICAL GROUP (AIChE)

(Programs in this group are accredited according to the program criteria for Chemical Engineering Technology and similarly named programs.)

Chemical Engineering Technology

Broome Community College ad
Memphis, State Technical Institute at ade
Pellissippi State Technical Community College ade
Thames Valley State Technical College ade
Trident Technical College ad
Wake Technical College ad*
Waterbury State Technical College ad

Chemical Technology

Cincinnati, University of; OMI College of Applied Science ad

CIVIL AND CONSTRUCTION GROUP (ASCE)

(Programs in this group are accredited according to the program criteria for Civil and Construction Engineering Technology and similarly named programs.)

Architectural and Building Engineering Technology

Vermont Technical College ad

Architectural and Civil Engineering Technology

Central Maine Vocational Technical Institute ad

Architectural Construction Option of Civil Engineering Technology

Owens Technical College ade

Building Energy Systems Technology

Pennsylvania State University, Fayette Campus ad

Civil and Construction Engineering Technology

Cincinnati, University of; OMI College of Applied Science ad

Civil Construction Technology

Stark Technical College ade

Civil Engineering Technology

Bluefield State College ad
Broome Community College ad
Central Piedmont Community College ade
Cincinnati Technical College ad
District of Columbia, University of the; Van Ness Campus ade
Erie Community College, North Campus ade
Fayetteville Technical Community College ad
Florence-Darlington Technical College ad
Franklin Institute of Boston ad
Gaston College ad
Guilford Technical Community College ad
Hartford State Technical College ade
Hawkeye Institute of Technology ad
Hudson Valley Community College ad
Indiana University-Purdue University at Fort Wayne ade
Indiana University-Purdue University at Indianapolis ade
Kansas College of Technology ad
Lowell, University of ae
Maine at Orono, University of ad
Michigan Technological University ad
Midlands Technical College ad
Murray State University ad*
Nashville State Technical Institute ade
Nassau Community College ade
New Mexico State University ad
New York City Technical College ade
New York, State University of, College of Technology, Canton ad
New York, State University of, College of Technology, Farmingdale ade
Norwalk State Technical College ade
Pellissippi State Technical Community College ade
Purdue University Calumet ade
Rochester Community College ad
St. Louis Community College at Florissant Valley ade
Southern College of Technology ade
Spartanburg Technical College ad
Staten Island, College of ade
Sumter Area Technical College ad
Trident Technical College ade
Toledo, University of; Community & Technical College ade
Vermont Technical College ad
Wake Technical College ad*
Wentworth Institute of Technology ad
Williamsport Area Community College ad
Youngstown State University ade

Civil Engineering Technology/ Construction Technology

Iowa Western Community College ad

Civil Technology

Mohawk Valley Community College ade
Northwest Mississippi Junior College ad
Rochester Institute of Technology ad

Civil-Surveying Engineering Technology[1]

West Virginia Institute of Technology ad

Civil/Construction Engineering Technology

Memphis, State Technical Institute at ade
Middlesex County College ade

Construction Engineering Technology

Arkansas at Little Rock, University of ade
Lawrence Institute of Technology ae
Nebraska, University of (Omaha Campus) ad
New York, State University of, College of Technology, Alfred ad
New York, State University of, College of Technology, Canton ad

Construction-Architectural Engineering Technology

New York, State University of, College of Technology, Farmingdale ade

Construction/Civil Engineering Technology

Mercer County Community College ade

Land Development Option in Civil Engineering Technology

Trident Technical College ade*

Public Works Engineering Technology

Oregon Institute of Technology ad.

Structural Engineering Technology

Oregon Institute of Technology ad

[1]Joint with ACSM under Surveying Group
*See note under Part I regarding period of accreditation.

Surveying and Construction Technology[1]

(See listing under Surveying Group)

COMPUTER GROUP (IEEE Lead Society, with IIE)

(Programs in this group are accredited according to the program criteria for Computer Engineering Technology and similarly named programs.)

Computer Engineering Technology

Capitol College ade
Central Piedmont Community College ade
District of Columbia, University of the; Van Ness Campus ade
Franklin Institute of Boston ad
Kansas College of Technology ad
Lake Superior State University ad
Memphis, State Technical Institute at ade
Murray State University ad*
Queensborough Community College ade
Santa Fe Community College ade
Savannah State College ad
Spring Garden College ade
Wake Technical College ade
Wentworth Institute of Technology ad

Computer Systems Engineering Technology

Oregon Institute of Technology ad

Computer Systems Option in Electrical/Electronic Engineering Technology

Chattanooga State Technical Community College ade

Computer Option in Electronics Engineering Technology

Amarillo College ad

Computer Electronics Option of Electrical/Electronics Engineering Technology

Owens Technical College ade

Computer Systems Technology

Norwalk State Technical College ade

Computer Technology

Hudson County Community College ade

[Computer-Aided Drafting and Design Option in Mechanical Engineering Technology]

[The program with this title is accredited according to the program criteria for Drafting/Design (Mechanical) Engineering Technology and Mechanical Engineering Technology]

[Computer-Aided-Manufacturing Technology]

[The program with this title is accredited according to the Program Criteria for Manufacturing Engineering Technology and Similarly Named Programs]

[Computing Graphics Engineering Technology]

[The program with this title is accredited only under the Drafting/Design (General) area]

Digital Logic Design Option in Electrical Engineering Technology

New York, State University of; College of Agriculture and Technology, Morrisville ad

DRAFTING/DESIGN (GENERAL) GROUP (TAC/ABET with ASCE, ASME & SME)

(Programs in this group are accredited according to the ABET general criteria for engineering technology programs.)

Computing Graphics Engineering Technology

New York, State University of, College of Technology, Alfred ad

Drafting/Design Engineering Technology

Nebraska, University of (Omaha Campus) ad

Engineering Graphics Technology

Piedmont Technical College ade

Engineering Design Graphics Technology

East Tennessee State University ad

DRAFTING/DESIGN (MECHANICAL) GROUP (ASME LEAD SOCIETY, with SME)

(Programs in this group are accredited according to the program criteria for Drafting/Design Engineering Technology (Mechanical) and similarly named programs.)

Computer-Aided-Drafting/Design Technology

Waterbury State Technical College ade

Computer Integrated Design and Drafting Technology

Pellissippi State Technical Community College ade

Design and Drafting Engineering Technology

Ricks College ad

Drafting and Design Engineering Technology

Lake Superior State University ad
West Virginia Institute of Technology ad

Drafting and Design Technology

Toledo, University of; Community and Technical College ade
Tri-State University ad

Design Engineering Technology

Stark Technical College ade

Engineering Graphics Technology

York Technical College ad

Industrial Drafting Technology

Rochester Institute of Technology ad

Mechanical Drafting and Design Engineering Technology

Forsyth Technical Community College ad

Mechanical Drafting and Design Technology

Guilford Technical Community College ad

Mechanical Drafting Design Technology

Indiana University-Purdue University at Fort Wayne ade
Indiana University-Purdue University at Indianapolis ade

Mechanical Design/Drafting with Computer-Aided-Design Option of Mechanical Engineering Technology

Owens Technical College ade
Owens Technical College (Findlay Campus) ade

ELECTRICAL AND ELECTRONIC GROUP (IEEE)

(Programs in this group are accredited according to the program criteria for Electrical/Electronic(s) Engineering Technology and similarly named programs.)

Automated Control and Instrumentation Option in Electrical/Electronic Engineering Technology

Chattanooga State Technical Community College ade

Biomedical Equipment Option of Electrical/Electronics Engineering Technology

(See listing under Bioengineering Technology Group)

Computer Systems Option in Electrical/Electronic Engineering Technology

(See listing under Computer Group)

Computer Electronics Option of Electrical/Electronics Engineering Technology

(See Listing under Computer (Group)

Computer Option in Electronics Engineering Technology

(See listing under Computer Group)

Digital Logic Design Option in Electrical Engineering Technology

(See listing under Computer Group)

[1]Joint with ACSM under Surveying Group

*See note under Part I regarding period of accreditation.

Electrical and Electronics Engineering Technology

- Vermont Technical College ade

Electrical Engineering Technology

- Bluefield State College ad
- Broome Community College ad
- Central Piedmont Community College ade
- Cincinnati, University of; OMI College of Applied Science ade
- Del Mar College ad
- Erie Community College, North Campus ade
- Gaston College ade
- Hartford State Technical College ade
- Hudson Valley Community College ad
- Indiana University-Purdue University at Fort Wayne ade
- Indiana University-Purdue University at Indianapolis ade
- Lawrence Institute of Technology ae
- Maine at Orono, University of ad
- Memphis, State Technical Institute at ade
- Michigan Technological University ad
- Middlesex County College ade
- Milwaukee School of Engineering ad
- Mohawk Valley Community College ade
- Murray State University ad*
- Nashville State Technical Institute ade
- New York City Technical College ade
- New York, State University of, College of Technology, Alfred ad
- New York, State University of, College of Technology, Canton ad
- Niagara County Community College ad
- Northeastern University ae
- Norwalk State Technical College ade
- Pennsylvania State University, Altoona Campus ad
- Pennsylvania State University, Beaver Campus ad
- Pennsylvania State University, Behrend College ad
- Pennsylvania State University, Berks Campus ad
- Pennsylvania State University, Delaware County Campus ad
- Pennsylvania State University, DuBois Campus ad
- Pennsylvania State University, Fayette Campus ad
- Pennsylvania State University, Hazleton Campus ad
- Pennsylvania State University, McKeesport Campus ad
- Pennsylvania State University, New Kensington Campus ad
- Pennsylvania State University, Ogontz Campus ad
- Pennsylvania State University, Schuylkill Campus ad
- Pennsylvania State University, Shenango Valley Campus ad
- Pennsylvania State University, Wilkes-Barre Campus ad
- Pennsylvania State University, Worthington-Scranton Campus ad
- Pennsylvania State University, York Campus ad
- Purdue University Calumet ade
- Purdue University Kokomo ade
- Purdue University North Central ade
- Purdue University (West Lafayette) ad
- St. Louis Community College at Florissant Valley ade
- Southern College of Technology ade
- Stark Technical College ade
- Staten Island, College of ade
- Thames Valley State Technical College ade
- Trident Technical College ade*
- Waterbury State Technical College ade
- Wentworth Institute of Technology ad
- West Virginia Institute of Technology ad
- Youngstown State University ade

Electrical Engineering Technology-Electronics

- Monroe Community College ade
- New York, State University of, College of Technology, Farmingdale ade
- Orange County Community College ade

Electrical Technology

- Bronx Community College ade

Electrical/Electronic(s) Engineering Technology

- Kent State University, Tuscarawas Campus ade
- Midlands Technical College ad
- Muskingum Area Technical College ade
- San Francisco, City College of ade
- Virginia Western Community College ad

Electronic(s) Engineering Technology

- Alamance Community College ad
- Amarillo College (General Electronics Option) ad
- Arkansas at Little Rock, University of ade
- Athens Area Technical Institute ade
- Blue Mountain Community College ad
- Broward Community College ade
- Broward Community College (North Campus) ade
- Capitol College ade
- Catawba Valley Technical College ade
- Central Piedmont Community College ade
- Champlain College ade
- Cogswell College, Cupertino Campus ad
- Cogswell College North ae
- Colorado Technical College, Colorado Springs Campus ade
- Columbus State Community College ade
- Davidson County Community College ad
- Davidson County Community College ae
- Dekalb Technical Institute ad
- Delta College ade
- DeVry Technical Institute, Woodbridge ad
- District of Columbia, University of the; Van Ness Campus ade
- Fairmont State College ad
- Fayetteville Technical Community College ad
- Florence-Darlington Technical College ade
- Forsyth Technical Community College ade
- Franklin Institute of Boston ad
- Franklin University ade
- Gaston College ade
- Glendale Community College ade
- Greenville Technical College ade
- Guilford Technical Community College ad
- Harrisburg Area Community College ade
- Hartford, University of; Samuel I. Ward College of Technology ade
- Horry-Georgetown Technical College ad
- Houston Community College System, Technical Education Center (General Electronics Option) ade
- Hudson County Community College ade
- Jefferson State Junior College ade
- John Tyler Community College ade
- Kansas College of Technology ad
- Lake Superior State University ad
- Lowell, University of ae
- Mercer County Community College ade
- Morris, County College of ade
- Nashville State Technical Institute ade
- Nebraska, University of (Omaha Campus) ad
- New Hampshire Technical Institute ade
- New Mexico State University ad
- Ocean County College ade
- Oregon Institute of Technology ad
- Parkland College ade
- Piedmont Technical College ade
- Portland Community College ade
- Prince George's Community College ade
- Queensborough Community College ade
- Ricks College ad
- Rochester Community College ad
- St. Louis Community College at Florissant Valley ade
- St. Petersburg Junior College ad
- Savannah Area Vocational-Technical School ad
- Sinclair Community College ade
- Spartanburg Technical College ad
- Spring Garden College ade
- Stark Technical College ade
- Technical Career Institutes ad
- Toledo, University of; Community & Technical College ade
- Tri-Cities State Technical Institute ade
- Tri-County Technical College ade
- Trident Technical College ade
- Triton College ade
- Wake Technical College ad
- Wentworth Institute of Technology ad
- West Virginia State College ade
- York Technical College ad

Electronic(s) Option in Electrical Engineering Technology

- New York, State University of; College of A&T, Morrisville ad
- Pellissippi State Technical Community College ade

Electronics Option of Electrical/Electronics Engineering Technology

- Owens Technical College ade

*See note under Part I regarding period of accreditation.

Electronics Option in Aeronautical Engineering Technology[a]

College of Aeronautics ade

Electronic(s) Technology

Akron, University of ade
Atlantic Community College ade
Montgomery College ade
Utah Valley Community College ade
Weber State College ade
Wentworth Institute of Technology ad
Wentworth Institute of Technology ae

Electro-Mechanical Option of Electrical/Electronics Engineering Technology

(See listing under Electromechanical Group)

Laser Electro-Optics Option in Electrical Engineering Technology

Hudson Valley Community College ad

Telecommunication Engineering Technology

Memphis, State Technical Institute at ade

Telecommunications Technology

Pennsylvania State University, Wilkes-Barre Campus ad

ELECTROMECHANICAL GROUP (IEEE Lead Society, with ASHRAE, ASME & SME)

(Programs in this group are accredited according to the program criteria for Electrical/Electronic(s) Engineering Technology and similarly named programs and the program criteria for Mechanical Engineering Technology and similarly named programs.)

Digital and Electromechanical Systems Engineering Technology

(See listing under Computer Group)

Electromechanical Engineering Technology

Dekalb Technical Institute ad
Michigan Technological University ad
New York City Technical College ade
Savannah Area Vocational-Technical School ad

Electro-Mechanical Engineering Technology

Cincinnati Technical College ade
New York, State University of, College of Technology, Alfred ad
Norwalk State Technical College ade
San Francisco, City College of ade
Staten Island, College of ade

[a]Joint with AIAA under Aerospace Group
[c]Joint with ASME under Mechanical Group
*See note under Part I regarding period of accreditation.

Electromechanical Technology

Rochester Institute of Technology ad*

Electrical-Mechanical Option of Electrical/Electronics Engineering Technology

Owens Technical College ade

INDUSTRIAL GROUP (IIE)

(Programs in this group are accredited according to the program criteria for Industrial Engineering Technology and similarly named programs.)

[Industrial Drafting Technology]

[The program with this title is accredited according to the program criteria for Drafting/Design (Mechanical) Engineering Technology only]

Industrial Engineering Technology

Catawba Valley Technical College ade
Gaston College ade
Indiana University-Purdue University at Fort Wayne ade
Lawrence Institute of Technology ae
Memphis, State Technical Institute at ade
Nashville State Technical Institute ade
Purdue University Calumet ade
Purdue University North Central ade
Southern College of Technology ade
Toledo, University of; Community & Technical College ade
Wake Technical College ade*

Industrial Management Engineering Technology

Staten Island, College of ade

INSTRUMENTATION GROUP (TAC of ABET with IEEE & ISA)

(Programs in this group are accredited according to the ABET general criteria for engineering technology programs.)

Instrumentation Engineering Technology

Tri-Cities State Technical Institute ade

Instrumentation Option in Electrical/Electronic Engineering Technology

(See listing under Electrical and Electronic Group)

MANUFACTURING GROUP (SME)

(Programs in this group are accredited according to the program criteria for Manufacturing Engineering Technology and similarly named programs.)

Computer-Aided-Manufacturing Option of Mechanical Engineering Technology[c]

Owens Technical College ade

Manufacturing Engineering Technology

Central Piedmont Community College ade
Cincinnati, University of, OMI College of Technology ae
Forsyth Technical Community College ad
Greater New Haven State Technical College ade
Hartford State Technical College ade
Nebraska, University of (Omaha Campus) ad
Oregon Institute of Technology ad
Ricks College ad
Thames Valley State Technical College ade
Waterbury State Technical College ade

Manufacturing Technology

Wentworth Institute of Technology ad

Manufacturing Option in Mechanical Engineering Technology[c]

Pellissippi State Technical Community College ade

MECHANICAL GROUP (ASME)

(Programs in this group are accredited according to the program criteria for Mechanical Engineering Technology and similarly named programs.)

Computer-Aided-Drafting and Design Option in Mechanical Engineering Technology

(See listing under Drafting/Design [Mechanical] Group)

Computer-Aided-Manufacturing Option of Mechanical Engineering Technology

(See listing under Manufacturing Group)

Internal Combustion Engines Option in Mechanical Engineering Technology

New York, State University of, College of Technology, Alfred ad

Manufacturing Option in Mechanical Engineering Technology

(See listing under Manufacturing Group)

Mechanical Design Engineering Technology

Michigan Technological University ad

Mechanical Design Technology

Wentworth Institute of Technology ae

[Mechanical Design/Drafting with Computer-Aided-Design Option of Mechanical Engineering Technology]

[Programs with this title are accredited according to the program criteria for Drafting/Design (Mechanical) Engineering Technology only]

[Mechanical Drafting and Design Engineering Technology]

[Programs with this title are accredited according to the program criteria for Drafting/ Design (Mechanical) Engineering Technology only]

[Mechanical Drafting and Design Technology]

[Programs with this title are accredited according to the program criteria for Drafting/ Design (Mechanical) Engineering Technology only]

[Mechanical Drafting Design Technology]

[Programs with this title are accredited according to the program criteria for Drafting/ Design (Mechanical) Engineering Technology only]

Mechanical Engineering Technology

- Arkansas at Little Rock, University of ade
- Bluefield State College ad
- Broome Community College ad
- Catawba Valley Technical College ade
- Central Piedmont Community College ade
- Chattanooga State Technical Community College (Mechanical Option) ade
- Cincinnati, University of; OMI College of Applied Science ade
- Cincinnati Technical College ade
- Cogswell College, Cupertino Campus ad
- Dekalb Technical Institute ad
- Delaware Technical Community College ade
- Delta College ade
- District of Columbia, University of the; Van Ness Campus ade
- Erie Community College, North Campus ade
- Franklin Institute of Boston ad
- Franklin University ade
- Gaston College ade
- Greater New Haven State Technical College ade
- Greenville Technical College ade
- Harrisburg Area Community College ade
- Hartford State Technical College ade
- Hudson Valley Community College ad
- Indiana University-Purdue University at Fort Wayne ade
- Indiana University-Purdue University at Indianapolis ade
- John Tyler Community College ade
- Kansas College of Technology ad
- Kent State University, Tuscarawas Campus ade
- Lake Superior State University ad
- Lawrence Institute of Technology ae
- Lowell, University of ae
- Maine at Orono, University of ad
- Memphis, State Technical Institute at ade
- Mercer County Community College ade
- Middlesex County College ade
- Midlands Technical College ad
- Milwaukee School of Engineering ad
- Morris, County College of ade
- Nashville State Technical Institute ade
- New Hampshire Technical Institute ade
- New Mexico State University ad
- New York City Technical College ade
- New York, State University of, College of Technology, Canton ad
- New York, State University of, College of Technology, Farmingdale ade
- New York, State University of, College of A&T, Morrisville ad
- Niagara County Community College ad
- Northeastern University ae
- Norwalk State Technical College ade
- Oregon Institute of Technology ad
- Pennsylvania State University, Altoona Campus ad
- Pennsylvania State University, Beaver Campus ad
- Pennsylvania State University, Behrend College ad
- Pennsylvania State University, Berks Campus ad
- Pennsylvania State University, DuBois Campus ad
- Pennsylvania State University, Hazleton Campus ad
- Pennsylvania State University, McKeesport Campus ad
- Pennsylvania State University, New Kensington Campus ad
- Pennsylvania State University, Ogontz Campus ad
- Pennsylvania State University, Shenango Valley Campus ad
- Pennsylvania State University, Wilkes-Barre Campus ad
- Pennsylvania State University, Worthington-Scranton Campus ad
- Pennsylvania State University, York Campus ad
- Purdue University Calumet ade
- Purdue University (West Lafayette) ad
- Purdue University at New Albany ade
- Queensborough Community College ade
- Rochester Community College ad
- St. Louis Community College at Florissant Valley ade
- San Francisco, City College of ade
- Sinclair Community College ade
- Southern College of Technology ade
- Spartanburg Technical College ad
- Spring Garden College ade
- Stark Technical College ade
- Staten Island, College of ade
- Thames Valley State Technical College ade
- Toledo, University of; Community & Technical College ade
- Tri-Cities State Technical Institute ade
- Trident Technical College ade
- Vermont Technical College ad
- Waterbury State Technical College ade
- Wentworth Institute of Technology ad
- West Virginia Institute of Technology ad
- Youngstown State University ade

Mechanical Technology

- Akron, University of ade
- Mohawk Valley Community College ade
- Wentworth Institute of Technology ad

Mechanical Option in Mechanical Engineering Technology

- Owens Technical College ade
- Pellissippi State Technical Community College ade

Product and Machine Design Option in Mechanical Engineering Technology

- New York, State University of, College of Technology, Alfred ad

MINING GROUP (SME-AIME)

(Programs in this group are accredited according to the program criteria for Mining Engineering Technology and Similarly named programs.)

Mining Engineering Technology

- Bluefield State College ad
- Fairmont State College ad
- Southern Indiana, University of ade*
- West Virginia Institute of Technology ad

NUCLEAR GROUP (ANS)

(Programs in this group are accredited according to the program criteria for Nuclear Engineering Technology and similarly named programs.)

Nuclear Engineering Technology

- Pennsylvania State University, Beaver Campus ad
- Thames Valley State Technical College ade

SURVEYING GROUP (ACSM Lead Society, with ASCE)

(Programs in this group are accredited according to the program criteria for Surveying Engineering Technology and similarly named programs.)

Civil-Surveying Engineering Technology[a]

(See listing under Civil and Construction Group)

Surveying and Construction Technology[a]

- Akron, University of ade

Surveying Engineering Technology

- Oregon Institute of Technology ad
- New York, State University of, College of Technology Alfred ad

[a]Joint with ASCE under Civil and Construction Group
*See note under Part I regarding period of accreditation.

Surveying Technology

East Tennessee State University ad

Mohawk Valley Community College ade

Pennsylvania State University, Wilkes-Barre Campus ad

WELDING GROUP (TMS Lead Society, with ASME & SME)

(Programs in this group are accredited according to the ABET general criteria for engineering technology programs.)

Welding Engineering Technology

Ricks College ad

OTHERS (ABET)

(Programs in this group are accredited according to the ABET general criteria for engineering technology programs.)

Apparel Engineering Technology

Southern College of Technology ade

Engineering Technology

Morrison Institute of Technology ad

Textile Engineering Technology

Southern College of Technology ade

Part II-B
Accredited Programs Leading to Bachelor's Degrees in Engineering Technology, 1989
By Program Area

The listings below are grouped according to the title of the program as reported by the institution and accredited by the Technology Accreditation Commission (TAC) of the Accreditation Board for Engineering and Technology (ABET). Options under overall program titles are listed with the major programs.

The letters in parentheses after the program titles indicate the Participating Bodies of ABET that have been assigned curricular responsibility for assisting ABET in the preparation of program criteria and the appointment of program evaluators for programs within their areas of professional disciplinary competence, as noted at the end of this listing. Lead societies have primary but not exclusive responsibility for assigned program areas. Where no society has been designated, ABET itself has curricular responsibility.

This list includes ONLY BACHELOR'S DEGREE PROGRAMS. For associate degree programs, refer to the preceding separate list.

Characteristics of the accredited programs are indicated after each program listing by combinations of the following symbols:

b—Baccalaureate program d—Day program e—Evening program de—Day and evening program

Note: Programs preceded by plus signs (+) are related to Associate Degree programs having the same program title.

AEROSPACE GROUP (AIAA)

(Programs in this group are accredited according to the ABET general criteria for engineering technology programs.)

Aeronautical Engineering Technology

Arizona State University bd

Aeronautical Operations Technology Option in Aeronautical Technology

New York Institute of Technology, Old Westbury Campus (jointly with College of Aeronautics) bde*

Aircraft Engineering Technology

Embry-Riddle Aeronautical University, Daytona Beach Campus bd

Aircraft Maintenance Engineering Technology

Northrop University bd

AGRICULTURAL GROUP (ASAE)

(Programs in this group are accredited according to the ABET general criteria for engineering technology programs.)

Agricultural Engineering Technology

Delaware, University of bd

AIR CONDITIONING GROUP (ASHRAE)

(Programs in this group are accredited according to the ABET general criteria for engineering technology programs.)

Air Conditioning and Refrigeration Technology

California Polytechnic State University, San Luis Obispo bd

*See note under Part I regarding period of accreditation.

+There is an Associate Degree program in the same curricular area at this institution.

ARCHITECTURAL GROUP (ASCE Lead Society, with ASHRAE)

(Programs in this group are accredited according to the ABET general criteria for engineering technology programs.)

Architectural and Building Construction Engineering Technology

(See listing under Civil and Construction Group)

Architectural Engineering Technology

+Cincinnati, University of; OMI College of Applied Science bde
+Southern College of Technology bde
Southern Mississippi, University of bd
+Wentworth Institute of Technology bd
+Wentworth Institute of Technology (Weekend) bw

Architectural Technology

Memphis State University bd

AUTOMOTIVE GROUP (SAE)

(Programs in this group are accredited according to the ABET general criteria for engineering technology programs.)

Automotive Engineering Technology

Weber State College bd

CIVIL AND CONSTRUCTION GROUP (ASCE)

(Programs in this group are accredited according to the program criteria for Civil and Construction Engineering Technology and similarly named programs.)

Civil and Construction Engineering Technology

Temple University bde

Civil Engineering Technology

Alabama A&M University bd
Alabama, University of bd
+Bluefield State College bde
Florida A&M University bd
Georgia Southern College bd
+Lowell, University of be
Metropolitan State College bde
North Carolina at Charlotte, University of bd
Northern Arizona University bd
Old Dominion University bd
Pittsburgh at Johnstown, University of bd
Point Park College bde
Rochester Institute of Technology bd
Savannah State College bd
South Carolina State College bd
+Southern College of Technology bde
+Southern Colorado, University of bd
Southern Illinois University at Carbondale bd
+Southern Indiana, University of bde
Tennessee at Martin, University of bd
+Wentworth Institute of Technology bd
Western Kentucky University bd
+Youngstown State University bde

Civil Engineering Technology Option in Engineering Technology

New Mexico State University bd

Construction and Contracting Option in Engineering Technology

New Jersey Institute of Technology bde
New Jersey Institute of Technology (Mercer County College) bde

Construction Engineering Technology

+Arkansas at Little Rock, University of bde
Central Connecticut State University bde
District of Columbia, University of the; Van Ness Campus bde
Louisiana Tech University bd
Montana State University bd
+Nebraska, University of (Omaha Campus) bd
Pittsburg State University bd
Southern Mississippi, University of bd

Construction Engineering Management Option in Engineering Technology

California State University, Sacramento bd

Construction Management Technology

Houston, University of; University Park bde
Oklahoma State University bd

Construction Technology

Akron, University of bde
East Tennessee State University bd
Indiana University-Purdue University at Fort Wayne bde
Murray State University bd
Purdue University Calumet bde
Texas Tech University bd

Construction Technology Option in Civil Engineering Technology

Oregon Institute of Technology bd

Public Works Option in Civil Engineering Technology

+ Oregon Institute of Technology bd

Structural Design and Construction Engineering Technology

Pennsylvania State University, Capital College bd

Structural Option in Civil Engineering Technology

+ Oregon Institute of Technology bd

COMPUTER GROUP (IEEE Lead Society, with IIE)

(Programs in this group are accredited according to the program criteria for Computer Engineering Technology and similarly named programs.)

[Computer Drafting and Design Technology]

[The program with this title is accredited according to the program criteria for Drafting/Design (Mechanical) Engineering Technology only]

Computer Engineering Technology

Arkansas at Little Rock, University of bde
+ Capitol College bde
Houston, University of, University Park bde
+ Murray State University bd
Southern Mississippi, University of bd
+ Spring Garden College bde
+ Wentworth Institute of Technology bd

Computer Engineering Technology Option in Engineering Technology

Kansas State University bd

+ There is an Associate Degree program in the same curricular area at this institution.

Computer Systems Engineering Technology

+ Oregon Institute of Technology bd

Computer Systems Technology

Memphis State University bd

Computer Technology

Rochester Institute of Technology bd
New York, State University of; College of Technology at Utica bde

Computer Technology Option in Engineering Technology

Central Florida, University of bd
Central Florida, University of (Brevard Campus) bd

Electromechanical Computer Technology

New York Institute of Technology, Metropolitan Center bde
New York Institute of Technology, Old Westbury Campus bde

Information Systems Technology Option in Engineering Technology

Central Florida, University of bd
Central Florida, University of (Brevard Campus) bd

DRAFTING/DESIGN (GENERAL) GROUP (TAC of ABET with ASCE, ASME & SME)

(Programs in this group are accredited according to the ABET general criteria for engineering technology programs.)

Drafting/Design Engineering Technology

+ Nebraska, University of (Omaha Campus) bd

DRAFTING/DESIGN (MECHANICAL) GROUP (ASME Lead Society, with SME)

(Programs in this group are accredited according to the program criteria for Drafting/Design (Mechanical) and similarly named programs.)

Computer Drafting and Design Technology

Houston, University of; University Park bde

Design Engineering Technology

Brigham Young University bd

Design Technology Option in Engineering Technology

Central Florida, University of bd
Central Florida, University of (Brevard Campus) bd

Mechanical Design/Drafting Engineering Technology

Pittsburg State University bd

Mechanical Drafting and Design Technology

Alabama A&M University bd

ELECTRICAL AND ELECTRONIC GROUP (IEEE)

(Programs in this group are accredited according to the program criteria for Electrical/Electronic(s) Engineering Technology and similarly named programs.)

Electrical and Electronic Engineering Technology

Montana State University bd

Electrical Engineering Technology

Alabama, University of bd
Alabama A&M University bd
+ Bluefield State College bd
Bradley University bd
+ Cincinnati, University of; OMI College of Applied Science bde
Georgia Southern College bd
+ Indiana University-Purdue University at Fort Wayne bde
Louisiana Tech University bd
+ Maine at Orono, University of; bd
+ Milwaukee School of Engineering bd
+ Murray State University bd
New Hampshire, University of bde
New York, State University of; College at Buffalo bde
New York, State University of; College of Technology, Utica bde
New York, State University of; College of Technology, Utica Extension at Hudson Valley Community College bde
New York at Binghamton, State University of bde
New York at Binghamton, State University of; Extension at Suny College of Technology at Alfred bde
North Carolina at Charlotte, University of bd
+ Northeastern University, bde
Northern Arizona University bd
Old Dominion University bd
Pennsylvania State University, Capital College bd
Pittsburgh at Johnstown, University of bd
Point Park College bde
+ Purdue University Calumet bde
+ Purdue University, Kokomo bde
Rochester Institute of Technology bde
Roger Williams College bde
Roger Williams College, Open Division bde
South Carolina State College bd
Southeastern Massachusetts University bd
+ Southern College of Technology bde
Southern Illinois University at Carbondale bd
+ Southern Indiana, University of bde
Temple University bde
Tennessee at Martin, University of bd
Western Kentucky University bd
+ Youngstown State University bde

Electrical Systems Option in Engineering Technology

New Jersey Institute of Technology bde
New Jersey Institute of Technology (Camden County College) bde

Electrical Technology

Houston, University of; University Park bde
+ Indiana University-Purdue University at Indianapolis bde
+ Purdue University (West Lafayette) bd
New York, State University of; College of Technology at Farmingdale bde

Electronic(s) Engineering Technology

Arizona State University bd
+ Arkansas at Little Rock, University of bde
Brigham Young University bd
+ Capitol College bde
Central Washington University bd
+ Cogswell College, Cupertino Campus bde
+ Cogswell College North be
+ Colorado Technical College, Colorado Springs Campus bde
Dayton, University of bde
DeVry Institute of Technology, Atlanta (Decatur) bd
DeVry Institute of Technology, Chicago bd
DeVry Institute of Technology, Columbus bd
DeVry Institute of Technology, Irving bd
DeVry Institute of Technology, Lombard bd
DeVry Institute of Technology, Los Angeles (City of Industry) bd
DeVry Institute of Technology, Kansas City bd
DeVry Institute of Technology, Phoenix bd
+ DeVry Technical Institute, Woodbridge bd
East Tennessee State University bde
+ Fairmont State College bd
Florida A&M University bd
Fort Valley State College bde
+ Franklin University bde
+ Hartford, University of; Samuel I. Ward College of Technology bde
+ Lake Superior State University bd
+ Lowell, University of be
Mankato State University bd
Metropolitan State College bde
+ Nebraska, University of (Omaha Campus) bd
+ Oregon Institute of Technology bd
Pittsburg State University bd
Savannah State College bd
+ Southern Colorado, University of bd
Southern Mississippi, University of bd
Southern Mississippi, University of-Gulf Park Campus be
+ Spring Garden College bde
Texas A&M University bd
Trenton State College bde
+ Weber State College bd
+ Wentworth Institute of Technology bd
+ West Virginia Institute of Technology bd
Western Washington University bd
Western Washington University (North Seattle Campus) bd

+ There is an Associate Degree program in the same curricular area at this institution.

Electronic(s) Engineering Technology Option in Engineering Technology

New Mexico State University bd

Electronic(s) Technology

+ Akron, University of bde
California Polytechnic State University, San Luis Obispo bd
Houston, University of; University Park bde
Memphis State University bd
+ Oklahoma State University bd

Electronic Technology Option in Engineering Technology

Central Florida, University of bd
Central Florida, University of (Brevard Campus) bd
Kansas State University bd

Electrical/Electronic(s) Technology

Texas Tech University bd

ELECTROMECHANICAL GROUP (IEEE Lead Society, with ASHRAE, ASME & SME)

(Programs in this group are accredited according to the program criteria for Electrical/Electronic(s) Engineering Technology and similarly named programs and the program criteria for Mechanical Engineering Technology and similarly named programs.)

Electromechanical Computer Technology

(See listing under Computer Group)

Electromechanical Technology

New York, City College of the City University of bde

Electro-Mechanical Engineering Technology

New York at Binghamton, State University of bde

ENERGY GROUP (TAC of ABET with ASME and IEEE)

(Programs in this group are accredited according to the ABET general criteria for engineering technology programs.)

Energy Engineering Technology

Rochester Institute of Technology bd

Energy Technology

Pennsylvania State University, Capital College bd

ENGINEERING TECHNOLOGY (GENERAL) GROUP (TAC of ABET with ASCE, ASHRAE, ASME & IEEE)

(Programs in this group are accredited according to the ABET general criteria for engineering technology programs.)

Engineering Technology

California State Polytechnic University, Pomona bd
[California State University, Sacramento—see Construction Management and Mechanical Technology options]
[Central Florida, University of—see Computer Technology, Design Technology, Electronics Technology, Information Systems Technology, and Operations Technology options]
[Houston, University of, Downtown College—see Process and Piping Design Engineering Technology option]

ENVIRONMENTAL AND SANITARY GROUP (AAEE Lead Society, with AIChE, ASCE, ASHRAE, ASME & SAE)

(Programs in this group are accredited according to the ABET general criteria for engineering technology programs.)

Engineering Technology (Environmental)

Norwich University bd

Environmental Engineering Technology

Pennsylvania State University, Capital College bd

INDUSTRIAL GROUP (IIE)

(Programs in this group are accredited according to the program criteria for Industrial Engineering Technology and similarly named programs.)

Industrial Engineering Technology

Dayton, University of bde
Georgia Southern College bd
+ Indiana University-Purdue University at Fort Wayne bde
New York, State University of, College of Technology, Utica bd
New York, State University of; College of Technology, Utica Extension at Hudson Valley Community College bde
+ Purdue University Calumet bde
+ Southern College of Technology bde
Southern Mississippi, University of bd
Southern Mississippi, University of-Gulf Park Campus be
Trenton State College bde

Industrial Engineering Technology Option in Engineering Technology

Kansas State University bd

MANUFACTURING GROUP (SME)

(Programs in this group are accredited according to the program criteria for Manufacturing Engineering Technology and similarly named programs.)

Manufacturing Engineering Technology

Arizona State University bd
Arkansas at Little Rock, University of bde
Brigham Young University bd
Central Connecticut State University bde
East Tennessee State University bd
Murray State University bd
+ Nebraska, University of (Omaha Campus) bd
+ Oregon Institute of Technology bd
Pittsburg State University bd
Rochester Institute of Technology bde
Texas A&M University bd
Weber State College bd
Wentworth Institute of Technology bd
Western Carolina University, bd
Western Washington University bd

Manufacturing Option in Mechanical Engineering Technology[6]

Indiana University-Purdue University at Fort Wayne bde

Manufacturing Processes Technology

California Polytechnic State University, San Luis Obispo bd

Manufacturing Systems Technology

Houston, University of; University Park bde

Manufacturing Technology

Memphis State University bd
Oklahoma State University bd
New York, State University of; College of Technology at Farmingdale bde

Manufacturing Technology Option in Engineering Technology

New Jersey Institute of Technology bde

Mechanical Design Option in Manufacturing Technology[6]

(See listing under Mechanical Group)

Production Operations Option in Manufacturing Technology

Bradley University bd

[6]Joint with ASME under Mechanical Group
[7]Joint with SME under Manufacturing Group
*See note under Part I regarding period of accreditation.
+ There is an Associate Degree program in the same curricular area at this institution.

MECHANICAL GROUP (ASME)

(Programs in this group are accredited according to the program criteria for Mechanical Engineering Technology and similarly named programs.)

Manufacturing Option in Mechanical Engineering Technology[7]

(See listing under Manufacturing Group)

Mechanical Design Option in Manufacturing Technology[7]

Bradley University bd

Mechanical Design Technology

Oklahoma State University bd

[Mechanical Design/Drafting Engineering Technology]

[The program with this title is accredited according to the program criteria for Drafting/Design (Mechanical) Engineering Technology only]

[Mechanical Drafting and Design Technology]

[The program with this title is accredited according to the program criteria for Drafting/Design (Mechanical) Engineering Technology only]

Mechanical Engineering Technology

Alabama A&M University bd
Arkansas at Little Rock, University of bde
+ Cincinnati, University of; OMI College of Applied Science bd
+ Cogswell College, Cupertino Campus bde
Dayton, University of bd
+ Franklin University bde
Georgia Southern College bd
+ Indiana University-Purdue University at Fort Wayne bde
+ Lake Superior State University bd
+ Lowell, University of be
+ Maine at Orono, University of bd
Metropolitan State College bde
+ Milwaukee School of Engineering bd
Montana State University bd
New Hampshire, University of bde
New York, State University of; College at Buffalo bde
New York, State University of; College of Technology, Utica bde
New York at Binghamton, State University of bde
North Carolina at Charlotte, University of bd
+ Northeastern University bde
Northern Arizona University bd
Old Dominion University bd
+ Oregon Institute of Technology bd
Pennsylvania State University, Capital College bd
Pittsburgh at Johnstown, University of bd
Point Park College bde
+ Purdue University Calumet bde
Rochester Institute of Technology bde
Roger Williams College bd*
Savannah State College bd
South Carolina State College bd
Southeastern Massachusetts University bd
+ Southern College of Technology bde
+ Southern Colorado, University of bd
Southern Illinois University at Carbondale bd
+ Southern Indiana, University of bde
Southern Mississippi, University of bd
+ Spring Garden College bde
Temple University bde
Tennessee at Martin, University of bd
Texas A&M University bd
Trenton State College bde
+ Wentworth Institute of Technology bd
Western Kentucky University bd
+ Youngstown State University bde

Mechanical Engineering Technology Option in Engineering Technology

California State University, Sacramento bd
Kansas State University bd
New Mexico State University bd

Mechanical Power Technology

Oklahoma State University bd

Mechanical Systems Option in Engineering Technology

New Jersey Institute of Technology bde

Mechanical Technology

+ Akron, University of bde
California Polytechnic State University, San Luis Obispo bd
+ Indiana University-Purdue University at Indianapolis bde
+ Purdue University (West Lafayette) bd
Texas Tech University bd

Mechanical/Structural Engineering Technology

Northeastern University be

MINING GROUP (SME-AIME)

(Programs in this group are accredited according to the ABET program criteria for Mining Engineering Technology and similarly named programs.)

Mining Engineering Technology

+ Bluefield State College bd
+ Fairmont State College bd
+ Southern Indiana, University of bde*
+ West Virginia Institute of Technology bd

NAVAL ARCHITECTURE AND MARINE GROUP (SNAME)

(Programs in this group are accredited according to the ABET general criteria for engineering technology programs.)

Marine Engineering Technology

California Maritime Academy bd
Maine Maritime Academy bd

PETROLEUM GROUP (SPE)

(Programs in this group are accredited according to the ABET general criteria for engineering technology programs.)

Petroleum Technology

Oklahoma State University bd

PLASTICS GROUP (TAC of ABET with AIChE, SAE & SME)

(Programs in this group are accredited according to the ABET general criteria for engineering technology programs.)

Plastics Engineering Technology

Pittsburg State University bd

SURVEYING GROUP (ACSM Lead Society, with ASCE)

(Programs in this group are accredited according to the program criteria for Surveying Engineering Technology and similarly named programs.)

Surveying and Mapping Technology

Houston, University of, University Park bde

WELDING GROUP (TMS Lead Society, with ASME & SME)

(Programs in this group are accredited according to the ABET general criteria for engineering technology programs.)

Welding Technology

California Polytechnic State University, San Luis Obispo bd

OTHERS (ABET)

(Programs in this group are accredited according to the ABET general criteria for engineering technology programs.)

Apparel Engineering Technology

+ Southern College of Technology bde

Automated Systems Engineering Technology

Lake Superior State University bd

Fire Protection and Safety Technology

Oklahoma State University bd

Operations Technology Option in Engineering Technology

Central Florida, University of bd
Central Florida, University of (Brevard Campus) bd

Process and Piping Design Engineering Technology Option in Engineering Technology

Houston, University of; Downtown College bde

Textile Engineering Technology

+ Southern College of Technology bde

Textile Management and Technology

Auburn University bd

+ There is an Associate Degree program in the same curricular area at this institution.

APPENDIX C

Examples of Engineering Curricula

RUTGERS UNIVERSITY

Four-Year Engineering Curricula

FRESHMAN-YEAR PROGRAM

Curriculum Code 004
(Common to all four-year curricula)

First Term		
01:160:159	General Chemistry for Engineers	3
01:160:171	Introduction to Experimentation*	1
01:350:101	Expository Writing *or* 14:440:127 Introduction to Computers for Engineers	3
14:440:100	Engineering Orientation Lectures	1
01:640:143	Calculus for Engineering	4
01:750:123	Analytical Physics I	2
—:—:—	Humanities/social sciences elective	3
Second Term		
01:160:160	General Chemistry for Engineers	3
14:440:127	Introduction to Computers for Engineers *or* 01:350:101 Expository Writing	3
01:640:144	Calculus for Engineering	4
14:655:221	Engineering Mechanics: Statics	3
01:750:124	Analytical Physics I	2
—:—:—	Humanities/social sciences elective	3
	Total Credits	35

APPLIED SCIENCES IN ENGINEERING

Four-Year Curriculum Code 073

	Freshman Year	
See Freshman Year Program		35
	Sophomore Year	
First Term		
01:640:243	Multivariable Calculus for Engineering	4
14:655:222	Engineering Mechanics: Dynamics	3
01:750:227	Analytical Physics IIA	3
01:750:229	Analytical Physics II Laboratory	1
—:—:—	Engineering or technical elective	3
—:—:—	Humanities/social sciences elective	3
Second Term		
01:220:200	Economic Principles and Problems	3
01:640:244	Differential Equations for Engineering and Physics	4
01:750:228	Analytical Physics IIB	3
01:750:230	Analytical Physics II Laboratory	1
—:—:—	Engineering or technical elective	3
—:—:—	Engineering or technical elective	3

*May be taken in the second term.

The last two years of the program must be developed with the assistance of the designated faculty adviser. The overall program must meet the student's career objectives and must be sufficiently different from the accredited engineering programs not to permit incorporation into an existing program. Applied sciences in engineering is not accredited as a professional engineering program.

The minimum degree credit requirement is 135. In addition to the specific courses shown above, the following distribution of courses must be completed:

- Humanities/social sciences electives: 12 credits
- Engineering electives: ten courses of 3 credits or more
- Mathematics/sciences electives: 6 credits
- General electives: 9 credits
- Technical electives: 27 credits

Some examples of possible concentrations in applied sciences in engineering are listed below. Many other fields may be covered to meet the special interests of engineering students. Specific courses are to be selected from those offered by the departments (see course descriptions at the end of this chapter). Availability of a particular course is not guaranteed.

Applied Mechanics. This concentration involves solid and fluid mechanics, elements of structures, and engineering analysis.

Biomedical Engineering. This concentration is intended to educate students in the application of analytical and experimental engineering techniques to biological and physiological problems.

Materials Science. Provision is made for the presentation of basic concepts of the structure of materials and the relation of that structure to physical properties.

Packaging Engineering. The packaging engineering concentration is designed to prepare engineers and scientists for a major role in the field of packaging. The program is structured to meet the technical requirements for the development and growth of total packaging systems.

Engineering Physics. This concentration allows students to combine a background in the basic engineering subjects with the courses of a physics curriculum. It provides a preparation for work in a physics research laboratory, further study in engineering, or graduate study in physics. The first two years are the same as those in any of the regular engineering curricula, although some substitutions are suggested. The last two years include courses in modern physics, electricity and magnetism, thermal physics, solid-state physics, and partial differential equations. Coupled with these are laboratory courses and other courses in engineering, physics, computer science, mathematics, or other sciences, to be chosen in consultation with an adviser in the Department of Physics.

BIORESOURCE ENGINEERING

Four-Year Curriculum Code 127

	Freshman Year	
See Freshman Year Program		35
	Sophomore Year	
First Term		
01:119:101	General Biology	4
14:440:117	Engineering Graphics	2
01:640:243	Multivariable Calculus for Engineering	4
14:655:222	Engineering Mechanics: Dynamics	3
01:750:227	Analytical Physics IIA	3
01:750:229	Analytical Physics II Laboratory	1
Second Term		
11:127:290	Biosystems Engineering Measurements	3
14:180:243	Mechanics of Solids	3
01:640:244	Differential Equations for Engineering and Physics	4
14:650:351	Elements of Thermodynamics	3
__:__:__	Option elective	3
	Junior Year	
First Term		
11:127:493	Unit Processes for Biological Materials	3
01:160:209	Elementary Organic Chemistry	3
01:160:211	Elementary Organic Chemistry Laboratory	1
14:180:387	Fluid Mechanics	3
14:180:389	Fluid Mechanics Laboratory	1
14:330:373	Elements of Electrical Engineering	3
14:330:375	Elements of Electrical Engineering Laboratory	1
__:__:__	General or option elective	3
Second Term		
11:127:492	Energy Conversion for Biological Systems	3
14:180:345	Properties of Materials Laboratory	1
14:330:374	Applied Electronics	3
__:__:__	Humanities/social sciences elective	3
__:__:__	Option elective	4
__:__:__	General elective	3
	Senior Year	
First Term		
11:127:488	Bioresource Engineering Design I	2
01:220:200	Economic Principles and Problems	3
__:__:__	Humanities/social sciences elective	3
__:__:__	Option elective	3
__:__:__	Option elective	3
__:__:__	Option elective	3
Second Term		
11:127:450	Applied Instrumentation and Control	3
11:127:489	Bioresource Engineering Design II	1
11:127:495	Environmental Systems Analysis for Engineers	3
__:__:__	General or option elective	3 or 4
__:__:__	Option elective	3
__:__:__	Option elective	3 or 4
	Total Credits	136–138

Students select one of the following options in order to develop an area of specialization. The unspecified option credits are selected with the approval of a student adviser. See the five-year Bioresource Engineering curriculum (129) presented later in this chapter for a listing of appropriate courses.

Food Engineering Option. 01:119:390; 01:160:320; 11:400:201, 402, and 411; 14:540:482 or 14:650:420; and 6 additional option credits. 14:155:303 and 308 may be substituted for 14:180:387, 389, and 14:650:351.

Bioenvironmental Engineering Option. 01:119:390; 11:127:494 and 496; 14:180:331 and 382; 11:375:444; 11:704:351; and 6 additional option credits.

Horticultural Engineering Option. 11:127:240, 490, and 491; 11:530:211 and 321; 11:930:266; and 6 additional option credits.

Agricultural Engineering Option. 01:119:102; 11:127:490, 491, and 494; 14:180:318 or 448 or 14:650:342; 11:930:251 or 266; and 6 additional option credits.

CERAMIC ENGINEERING

Four-Year Curriculum Code 150

The ceramic engineering curriculum includes three options: general, science, and technical management.

	Freshman Year	
See Freshman Year Program		35
	Sophomore Year	
(Common to all options except as noted)		
First Term		
14:150:203	Introductory Ceramics	3
01:220:200	Economic Principles and Problems	3
01:460:301	Mineralogy: Petrology I	4
01:640:243	Multivariable Calculus for Engineering	4
01:750:227	Analytical Physics IIA	3
01:750:229	Analytical Physics II Laboratory	1
Second Term		
14:150:204	Introductory Ceramics	2
14:150:254	Introductory Ceramics Laboratory	1
01:640:244	Differential Equations for Engineering and Physics	4
14:655:222	Engineering Mechanics: Dynamics*	3
01:750:228	Analytical Physics IIB	3
01:750:230	Analytical Physics II Laboratory	1
__:__:__	Humanities/social sciences elective	3

General Option

	Junior Year	
First Term		
14:150:301	Measurements in Physical Ceramics	4
14:150:303	Thermodynamics for Ceramics	3
14:150:305	Physics of Ceramics	3
14:150:307	Unit Operations and Processes I	3
14:150:309	Analytical Techniques for Ceramics	2
14:150:359	Analytical Techniques for Ceramics Laboratory	1
Second Term		
14:150:304	Ceramic Compositions	3
14:150:354	Ceramic Compositions Laboratory	1
14:150:306	Phase Diagrams for Ceramics	3
14:150:308	Unit Operations and Processes II	3
14:330:373	Elements of Electrical Engineering	3
14:330:375	Elements of Electrical Engineering Laboratory	1
__:__:__	Technical elective*	3
	Senior Year	
First Term		
14:150:403	Senior Ceramics Seminar	1
14:150:406	Refractories	3
14:150:411	Ceramic Engineering Design	3
14:150:415	Glass Engineering	3
14:150:455	Glass Engineering Laboratory	1
14:540:343	Engineering Economics	3
__:__:__	Humanities/social sciences elective	3
Second Term		
14:150:312	Ceramic Raw Materials	3
14:150:404	Senior Ceramics Seminar	1

*Science Option and Technical Management Option: 14:330:373 or 374 may be substituted for 14:655:222.

14:150:412	Ceramic Engineering Design	3
14:150:__	Departmental elective*	3
14:150:__	Departmental elective*	3
—:—:—	General elective	3
	Total Credits	136

Science Option

Junior Year

First Term		
14:150:301	Measurements in Physical Ceramics	4
14:150:303	Thermodynamics for Ceramics	3
14:150:305	Physics of Ceramics	3
14:150:309	Analytical Techniques for Ceramics	2
14:150:359	Analytical Techniques for Ceramics Laboratory	1
01:960:401	Basic Statistics for Research	3
Second Term		
14:150:304	Ceramic Compositions	3
14:150:354	Ceramic Compositions Laboratory	1
14:150:306	Phase Diagrams for Ceramics	3
14:150:407	Ceramic Microscopy	3
14:150:457	Ceramic Microscopy Laboratory	1
01:960:490	Introduction to Experimental Design	3
—:—:—	Technical elective*	3

Senior Year

First Term		
14:150:307	Unit Operations and Processes I	3
14:150:401	Ceramics Laboratory I	3
14:150:403	Senior Ceramics Seminar	1
14:150:__	Departmental elective*	3
14:150:__	Departmental elective*	3
—:—:—	Humanities/social sciences elective	3
Second Term		
14:150:308	Unit Operations and Processes II	3
14:150:402	Ceramics Laboratory II	3
14:150:404	Senior Ceramics Seminar	1
14:150:421	Electronic Ceramics	3
14:150:__	Departmental elective*	3
—:—:—	General elective	3
	Total Credits	135

Technical Management Option

Junior Year

First Term		
33:010:310	Accounting for Engineers	3
14:150:301	Measurements in Physical Ceramics	4
14:150:307	Unit Operations and Processes I	3
14:150:309	Analytical Techniques for Ceramics	2
14:150:359	Analytical Techniques for Ceramics Laboratory	1
01:960:401	Basic Statistics for Research	3
Second Term		
14:150:304	Ceramic Compositions	3
14:150:354	Ceramic Compositions Laboratory	1
14:150:308	Unit Operations and Processes II	3
14:540:343	Engineering Economics	3
14:150:__	Departmental elective*	3
—:—:—	Humanities/social sciences elective	3

Senior Year

First Term		
14:150:403	Senior Ceramics Seminar	1
14:150:411	Ceramic Engineering Design	3
14:150:413	Ceramic Engineering Venture Analysis	3
14:150:415	Glass Engineering	3
14:150:__	Departmental elective*	3
—:—:—	General elective	3
Second Term		
14:150:404	Senior Ceramics Seminar	1
14:150:312	Ceramic Raw Materials	3
14:150:412	Ceramic Engineering Design	3
33:630:301	Principles of Marketing	3
14:150:__	Departmental elective*	3
—:—:—	General elective	3
	Total Credits	134

*No more than one course numbered 14:150:37_, 47_, or 48_ may be used.

CHEMICAL ENGINEERING

Four-Year Curriculum Code 155

The chemical engineering curriculum includes two options: chemical and biochemical.

Freshman Year

See Freshman Year Program		35

Sophomore Year

(Common to both options)

First Term		
01:160:307	Organic Chemistry	4
01:220:200	Economic Principles and Problems	3
01:640:243	Multivariable Calculus for Engineering	4
14:655:222	Engineering Mechanics: Dynamics	3
01:750:227	Analytical Physics IIA	3
01:750:229	Analytical Physics II Laboratory	1
Second Term		
14:155:204	Chemical Engineering Analysis I	4
01:160:308	Organic Chemistry	4
01:640:244	Differential Equations for Engineering and Physics	4
01:750:228	Analytical Physics IIB	3
01:750:230	Analytical Physics II Laboratory	1
—:—:—	Humanities/social sciences elective	3

Chemical Option

Junior Year

First Term		
14:155:303	Transport Phenomena in Chemical Engineering I	4
14:155:307	Chemical Engineering Analysis II	3
01:160:311	Organic Chemistry Laboratory	2
01:160:323	Physical Chemistry	3
01:160:325	Physical Chemistry Laboratory for Engineers	2
—:—:—	General elective	3
Second Term		
14:155:304	Transport Phenomena in Chemical Engineering II	4
14:155:306	Chemical Engineering Measurements	2
14:155:308	Chemical Engineering Thermodynamics	4
01:160:324	Physical Chemistry	3
—:—:—	Humanities/social sciences elective	3
—:—:—	General elective	3

Senior Year

First Term		
14:155:409	Chemical Systems Safety and Health Engineering Management	1.5
14:155:415	Process Engineering I	3
14:155:441	Chemical Engineering Kinetics	3
14:155:423	Design of Separation Processes	3
14:330:373	Elements of Electrical Engineering	3
—:—:—	Technical elective	3
Second Term		
14:155:416	Process Engineering II	3
14:155:424	Chemical Engineering Design and Economics	3
14:155:422	Process Simulation and Control	3
—:—:—	Technical elective	3
	Total Credits	136.5

Biochemical Option

Junior Year

First Term

01:119:390	General Microbiology	4
14:155:303	Transport Phenomena in Chemical Engineering I	4
14:155:307	Chemical Engineering Analysis II	3
01:160:341	Physical Chemistry: Biochemical Systems	3
__:__:__	General elective	3

Second Term

01:115:301	Introductory Biochemistry	3
01:115:313	Introductory Biochemistry Laboratory	1
14:155:304	Transport Phenomena in Chemical Engineering II	4
14:155:306	Chemical Engineering Measurements	2
14:155:308	Chemical Engineering Thermodynamics	4
01:160:342	Physical Chemistry: Biochemical Systems	3
__:__:__	Humanities/social sciences elective	3

Senior Year

First Term

14:155:409	Chemical Systems Safety and Health Engineering Management	1.5
14:155:411	Introduction to Biochemical Engineering I	3
14:155:415	Process Engineering I	3
14:155:423	Design of Separation Processes	3
14:155:441	Chemical Engineering Kinetics	3
14:330:373	Elements of Electrical Engineering	3

Second Term

14:155:412	Introduction to Biochemical Engineering II	3
14:155:414	Biochemical Engineering Laboratory	3
14:155:422	Process Simulation and Control	3
14:155:424	Chemical Engineering Design and Economics	3
	Total Credits	137.5

Note: (Both options) (1) 14:330:373 may be taken in the junior year and the general or humanities/social sciences elective switched to the senior year with the approval of the student's adviser. (2) The technical elective is a junior- or senior-level mathematics, science, or engineering course selected with the approval of the student's adviser.

CIVIL ENGINEERING

Four-Year Curriculum Code 180

Freshman Year

See Freshman Year Program 35

Sophomore Year

First Term

14:180:209	Geometronics	4
01:220:200	Economic Principles and Problems	3
01:640:243	Multivariable Calculus for Engineering	4
14:655:222	Engineering Mechanics: Dynamics	3
01:750:227	Analytical Physics IIA	3
01:750:229	Analytical Physics II Laboratory	1

Second Term

14:180:243	Mechanics of Solids	3
14:440:117	Engineering Graphics	2
01:640:244	Differential Equations for Engineering and Physics	4
01:750:228	Analytical Physics IIB	3
01:750:230	Analytical Physics II Laboratory	1
__:__:__	Humanities/social sciences elective	3

Junior Year

First Term

14:180:301	Civil Engineering Analysis	3
14:180:345	Properties of Materials Laboratory	1
14:180:387	Fluid Mechanics	3
14:180:389	Fluid Mechanics Laboratory	1
14:180:__	Departmental elective	3
14:655:407	Mechanical Properties of Materials	3
__:__:__	Humanities/social sciences elective	3

Second Term

14:180:318	Elements of Structures	3
14:180:320	Elements of Structural Design	3
14:180:372	Soil Mechanics	3
14:180:374	Soil Mechanics Laboratory	1
14:180:__	Departmental elective	3
__:__:__	Technical elective	3

Senior Year

First Term

14:180:411	Reinforced Concrete	3
14:180:421	Reinforced Concrete Laboratory	1
14:180:473	Foundation Engineering	2
14:180:475	Foundation Engineering Laboratory	1
14:180:__	Departmental elective	3
__:__:__	Technical elective	3
__:__:__	General elective	3

Second Term

14:180:__	Departmental elective (Capstone Design)	3 or 4
14:180:__	Departmental elective	3
14:180:__	Departmental elective	3
__:__:__	Technical elective	3
__:__:__	Technical elective	3
	Total Credits	133 or 134

1. Departmental courses may be used for technical electives, but technical courses outside the department may not be used as departmental electives. The department publishes annually a list of acceptable technical electives. The following technical courses are strongly recommended:
   - 14:330:373 Elements of Electrical Engineering
   - 14:540:343 Engineering Economics
   - 14:650:351 Thermodynamics
2. At least one of the following Capstone Design courses is required:
   - 14:180:407 Construction Projects
   - 14:180:426 Structural Design
   - 14:180:431 Design of Environmental Engineering Facilities
   - 14:180:474 Geotechnical Engineering
3. With a combination of the required and elective departmental courses and the available general and technical electives, a student may pursue a general program or a program having an area of concentration in structural engineering, geotechnical engineering, construction engineering, or water resources/environmental engineering.

ELECTRICAL ENGINEERING

Four-Year Curriculum Code 330

The electrical engineering curriculum includes two options: electrical engineering and computer engineering.

Freshman Year

See Freshman Year Program 35

Sophomore Year

(Common to both options)

First Term

14:330:221	Principles of Electrical Engineering I	3
14:330:223	Principles of Electrical Engineering I Laboratory	1

14:330:225	Computer Organization and Logic Design	3
16:640:243	Multivariable Calculus for Engineers	4
01:750:227	Analytical Physics IIA	3
01:750:229	Analytical Physics II Laboratory	1
—:—:—	Humanities/social sciences elective	3
Second Term		
01:220:200	Economic Principles and Problems	3
14:330:222	Principles of Electrical Engineering II	3
14:330:224	Principles of Electrical Engineering II Laboratory	1
14:330:227	Programming Methodology and Data Structures	3
01:640:244	Differential Equations for Engineering and Physics	4
01:750:228	Analytical Physics IIB	3
01:750:230	Analytical Physics II Laboratory	1

Electrical Engineering Option

Junior Year

First Term		
14:330:345	Linear Systems and Signals	3
14:330:347	Linear Systems and Signals Laboratory	1
14:330:349	Probability and Stochastic Processes	4
14:330:361	Electronic Devices	3
14:330:363	Electronic Devices Laboratory	1
14:330:403	Electrical Energy Conversion	3
14:330:405	Electrical Energy Conversion Laboratory	1
—:—:—	Humanities/social science elective*	3
Second Term		
14:330:335	Electromagnetic Fields	3
14:330:346	Digital Signal Processing	3
14:330:348	Digital Signal Processing Laboratory	1
14:330:362	Analog Electronics	3
14:330:364	Analog Electronics Laboratory	1
—:—:—	Humanities/social science elective	3
—:—:—	General elective	3

Senior Year

First Term		
14:330:404	Introduction to Automatic Control Theory	3
14:330:450	Principles of Communication Systems	3
14:330:—	Technical elective	3
14:330:—	Technical elective	3
—:—:—	Technical elective	3
Second Term		
14:330:415	Digital Electronics	3
14:330:417	Digital Electronics Laboratory	1
14:330:—	Technical elective	3
14:330:—	Technical elective	3
—:—:—	Technical elective	3
—:—:—	Technical elective	3
	Total Credits	138

Computer Engineering Option

Junior Year

First Term		
14:330:337	Computer Architecture and Assembly Language	3
14:330:339	Computer Architecture Laboratory	1
14:330:345	Linear Systems and Signals	3
14:330:347	Linear Systems and Signals Laboratory	1
14:330:361	Electronic Devices	3
14:330:363	Electronic Devices Laboratory	1
—:—:—	Technical elective	3
—:—:—	Humanities/social science elective*	3

*Both options require an additional 3 credits of upper-level, humanities/social science elective credit beyond the minimum college requirements.

Second Term		
01:198:315	Language Software	4
14:330:335	Electromagnetic Fields	3
14:330:346	Digital Signal Processing	3
14:330:348	Digital Signal Processing Laboratory	1
14:330:415	Digital Electronics	3
14:330:417	Digital Electronics Laboratory	1
—:—:—	Humanities/social science elective	3

Senior Year

First Term		
01:198:416	Operating Systems	4
14:330:450	Principles of Communication Systems	3
—:—:—	Computer elective	3
—:—:—	Technical elective	3
—:—:—	Technical elective	3
Second Term		
14:330:472	Introduction to Software Engineering	3
—:—:—	Computer elective	3
—:—:—	Computer elective	3
—:—:—	Technical elective	3
—:—:—	General elective	3
	Total Credits	138

Note: A list of technical and computer electives is published by the department. These courses must be selected with the approval of the departmental adviser.

INDUSTRIAL ENGINEERING

Four-Year Curriculum Code 540

Freshman Year

See Freshman Year Program 35

Sophomore Year

First Term		
14:440:115	Basic Engineering Drawing	1
14:540:201	Work Design and Productivity	3
14:540:202	Work Design and Productivity Laboratory	1
14:540:210	Engineering Probability	3
01:640:243	Multivariable Calculus for Engineering	4
01:750:227	Analytical Physics IIA	3
01:750:229	Analytical Physics II Laboratory	1
Second Term		
01:350:302	Scientific and Technical Writing	3
14:540:203	Manufacturing Processes	3
14:540:204	Manufacturing Processes Laboratory	1
01:640:244	Differential Equations for Engineering and Physics	4
14:655:222	Engineering Mechanics: Dynamics	3
01:750:228	Analytical Physics IIB	3
01:750:230	Analytical Physics II Laboratory	1

Junior Year

First Term		
33:010:310	Accounting for Engineers	3
14:180:343	Mechanics of Solids	3
01:220:200	Economic Principles and Problems	3
14:330:373	Elements of Electrical Engineering	3
14:540:311	Deterministic Models in Operations Research	3
01:960:380	Intermediate Statistical Analysis	3
Second Term		
14:540:313	Operations Research Laboratory	1
14:540:333	Quality Control	3
14:540:338	Probability Models in Operations Research	3
14:540:343	Engineering Economics	3
14:650:420	Machine Design	3
—:—:—	Humanities/social sciences elective	3

Senior Year

First Term

14:540:421	Industrial Organization and Management	3
14:540:453	Production Planning and Control	3
14:540:462	Facilities Layout and Materials Handling	3
14:540:482	Computer Control of Manufacturing Systems	3
14:540:483	Computer Control of Manufacturing Systems Laboratory	1
__:__:__	Departmental or technical elective	3
__:__:__	Humanities/social sciences elective	3

Second Term

14:540:481	Simulation Models in Industrial Engineering	3
14:540:484	Design of Manufacturing Projects	3
14:650:351	Elements of Thermodynamics	3
__:__:__	Departmental or technical elective	3
__:__:__	General elective	3
	Total Credits	137

MECHANICAL ENGINEERING

Four-Year Curriculum Code 650

The mechanical engineering curriculum includes two options: mechanical engineering and aerospace engineering.

Freshman Year

See Freshman Year Program 35

Sophomore Year

First Term

01:220:200	Economic Principles and Problems	3
01:640:243	Multivariable Calculus for Engineering	4
14:650:231	Mechanical Engineering Computational Analysis and Design *or* 14:650:282 Thermodynamics	3
14:655:222	Engineering Mechanics: Dynamics	3
01:750:227	Analytical Physics IIA	3
01:750:229	Analytical Physics II Laboratory	1

Second Term

14:330:373	Elements of Electrical Engineering	3
14:330:375	Elements of Electrical Engineering Laboratory	1
14:440:115	Basic Engineering Drawing	1
01:640:244	Differential Equations for Engineering and Physics	4
14:650:282	Thermodynamics *or* 14:650:231 Mechanical Engineering Computational Analysis and Design	3
01:750:228	Analytical Physics IIB	3
01:750:230	Analytical Physics II Laboratory	1

Mechanical Engineering Option

Junior Year

First Term

14:540:343	Engineering Economics	3
01:640:421	Advanced Calculus for Engineering	3
14:650:311	Continuum Mechanics I	3
14:650:350	Mechanical Engineering Measurements *or* 14:650:312 Fluid Mechanics	4
__:__:__	Humanities/social sciences elective	3

Second Term

14:650:302	Applied Thermodynamics	3
14:650:312	Fluid Mechanics *or* 14:650:350 Mechanical Engineering Measurements	3
14:650:342	Design of Mechanical Components	3
14:655:407	Mechanical Properties of Materials	3
__:__:__	Technical elective	3

Senior Year

First Term

14:650:431	Mechanical Engineering Laboratory I	1
14:650:481	Heat Transfer*	3
14:650:487	Design of Mechanical Systems*	3
14:650:488	Computer Aided Design*	3
14:650:__	Departmental elective	3
14:650:__	Departmental elective	3

Second Term

14:650:432	Mechanical Engineering Laboratory II	1
14:650:443	Vibrations and Controls*	3
14:650:__	Departmental elective	3
__:__:__	General elective	3
__:__:__	Technical elective	3
__:__:__	Humanities/social sciences elective	3
	Total Credits	131

Aerospace Engineering Option

Junior Year

First Term

14:540:343	Engineering Economics	3
01:640:421	Advanced Calculus for Engineering	3
14:650:311	Continuum Mechanics I	3
14:650:350	Mechanical Engineering Measurements *or* 14:650:312 Fluid Mechanics	4
__:__:__	Humanities/social sciences elective	3

Second Term

14:650:302	Applied Thermodynamics	3
14:650:312	Fluid Mechanics *or* 14:650:350 Mechanical Engineering Measurements	3
14:650:342	Design of Mechanical Components	3
14:650:458	Aerospace Structures	3
14:655:407	Mechanical Properties of Materials	3

Senior Year

First Term

14:650:431	Mechanical Engineering Laboratory I	1
14:650:459	Aerospace Propulsion	3
14:650:460	Aerodynamics	3
14:650:481	Heat Transfer*	3
14:650:487	Design of Mechanical Systems*	3
14:650:__	Departmental elective	3

Second Term

14:650:433	Aerospace Engineering Laboratory	1
14:650:443	Vibrations and Controls*	3
14:650:488	Computer Aided Design*	3
__:__:__	General elective	3
__:__:__	Technical elective	3
__:__:__	Humanities/social sciences elective	3
	Total Credits	131

Beginning with the class of 1992, 14:650:311 Continuum Mechanics will be taken in the sophomore year in place of 14:650:282 Thermodynamics, and only one thermodynamics course will be required in the curriculum. Contact the mechanical engineering academic adviser or executive officer for further information.

*May be taken during the first or second term.

IOWA STATE UNIVERSITY

College of Engineering

Engineers occupy a uniquely important position in our modern civilization. They have the responsibility for taking the discoveries of basic science and translating them into products, structures, facilities, and services for the use of society.

Objectives of Curricula in Engineering

The broad objectives of engineering education are to develop professional competence and, by breadth of study, to prepare students for participation as leaders in the affairs of their professions, their communities, the state, and the nation. Engineering education seeks to develop a capacity for objective and analytical thought.

The curricula in engineering provide a balanced program in mathematics, basic sciences, engineering sciences, engineering design, and social sciences and humanities (SSH). This content is consistent with requirements of the Engineering Accreditation Commission of the Accreditation Board for Engineering and Technology (EAC/ABET), the national engineering accrediting agency.

Registration as a professional engineer, which is granted by the individual states, is required for many types of positions. The professional curricula in engineering at Iowa State University are designed to prepare a graduate for subsequent registration in all states. Seniors in EAC/ABET accredited curricula of the College of Engineering are encouraged to take the Fundamentals Examination for professional registration during their final academic year. Seniors in engineering curricula who have obtained at least 14 semester credits in surveying may take the Fundamentals Examination for professional registration as land surveyors.

Advanced work in engineering is offered in the post-graduate programs. See the *Graduate College* section of this catalog.

Organization of Curricula

All curricula in engineering are divided into two phases: a basic program and a professional program. The basic program consists primarily of subjects fundamental and common to all branches of engineering and includes chemistry, physics, mathematics, engineering graphics, engineering computations, and English. A student who has adequate high school preparation is expected to complete the basic program in one year. The professional phase of a curriculum includes intensive study in the particular branch of engineering which a student chooses, as well as a continuation of supporting work in mathematics, basic sciences, humanities, and social sciences.

Students must complete the requirements of the basic program before proceeding to a professional program. Prior to admission to a professional program, students will be considered to be in a preprofessional program.

*Students planning to enroll in Ch E, Cer E, or Met E will find 177 to be a better preparation for Chem 178. However, Chem 167 is accepted as a substitute for 177 for those students declaring one of these curricula late.

Curricula and Administering Departments*

Curriculum	Administering department	See key below					
Aerospace Engr.	Aerospace Engr.	A	B	C	D	E	F
Agricultural Engr.	Agricultural Engr.	A	B	C	D	E	F
Biomedical Engr.	Multi-departmental					E	F
Ceramic Engr.	Materials Science and Engr	A		C		E	F
Chemical Engr.	Chemical Engr.	A	B	C	D	E	F
Civil Engr.	Civil & Construction Engr.	A	B	C		E	
Civil Engr. Materials	Civil & Construction Engr.					E	F
Computer Engr.	Elec. Engr. and Comp. Engr.	A	B	C	D	E	F
Construction Engr.	Civil & Construction Engr.	A	B	C			
Electrical Engr.	Elec. Engr. and Comp. Engr.	A	B	C	D	E	F
Engr. Journalism	Industrial Engr.	A					
Engr. Mechanics	Engr. Science and Mech.				D	E	F
Engr. Operations	Industrial Engr.	A	B				
Engr. Science	Engr. Science and Mech.	A	B	C			
Engr. Valuation	Industrial Engineering				D	E	F
Geodesy & Photogram.	Civil & Construction Engr.					E	
Geotechnical Engr.	Civil & Construction Engr.					E	F
Industrial Engr.	Industrial Engr.	A	B	C	D	E	F
Materials Sci. & Engr.	Materials Science & Engr.				D		
Mechanical Engr.	Mechanical Engr.	A	B	C		E	F
Metallurgical Engr.	Materials Science and Engr.	A	B	C			
Metallurgy	Materials Science and Engr.					E	F
Municipal Engr.	Civil & Construction Engr.					E	
Nuclear Engr.	Nuclear Engr.	A	B	C	D	E	F
Operations Research	Industrial Engr.					E	F
Sanitary Engr.	Civil & Construction Engr.					E	F
Structural Engr.	Civil & Construction Engr.					E	F
Transportation Engr.	Civil & Construction Engr.					E	F

Key:

A Offers 4 year bachelor of science (B.S.) degree
B Offers 5 year bachelor of science degree (co-op program)
C Undergraduate program accredited by EAC/ABET
D Offers master of engineering (M. Engr.) degree
E Offers master of science (M.S.) degree
F Offers doctor of philosophy (Ph.D.) degree
***** The college also has a Department of Freshman Engineering (see page 169).

Preparation for the Engineering Curricula

Students who wish to study engineering should have high school credits for 2 years of algebra, 1 year of geometry, and ½ year of trigonometry; 1 year each of chemistry and physics, and 4 years of English. Placement examinations in mathematics will be given to all new students. Advanced placement is possible for students with exceptional high school preparation. Students who score poorly on these examinations may be encouraged or required to take remedial courses and should expect to spend more than the customary time to complete the engineering program.

Basic Program for Professional Engineering Curricula

The first year program is much the same for all professional curricula in the College of Engineering. Each curriculum requires completion of the basic program as well as the curriculum designated requirements. The basic program is a set of core courses common to all engineering curricula, while the curriculum designated requirements are courses required by individual curricula. The student who desires to receive the bachelor's degree in a minimum time will find it desirable to select a curriculum as soon as possible.

Entering undergraduates must demonstrate proficiency in trigonometry on the mathematics placement examination or have passed a college trigonometry course before enrolling in Math 166. Students who are not adequately prepared may have to take Math 140, 141, or 142 and/or Chem 50 in addition to the courses listed below. Neither Math 140, 141, 142 nor Chem 50 may be used to satisfy credit requirements for graduation in any of the engineering curricula.

Basic Program

Cr.	
8 or 10	Mathematics 165, 166 or 175, 176
6	English 104, 105
6	Freshman Engineering 160, 170
4	Chemistry 167 or 177*
5	Physics 221
R	Freshman Engineering 101
0.5	Library 160
29.5-31.5	**Total credits**

Curriculum Designated Requirements

Aerospace Engineering—SSH electives (6 cr.)

Agricultural Engineering—Chem 167L, A E 110, Agron 154 or F Tch 101 (FE) or B B 221 (PE)

Ceramic Engineering—Chem 177*, 177L, SSH elective (3 cr.)

Chemical Engineering—Chem 177*, 177L, 178, 178L, Com S 205

Civil Engineering—Chem 167L, a course in statistics from a department-approved list (2 cr.)

Computer Engineering—Com S 211, 212

Construction Engineering—Con E 110, Psych 101

Electrical Engineering—Com S 211, SSH elective (3 cr.)

Reprinted courtesy of Iowa State University.

Engineering Operations—Psych 101
Engineering Science—Chem 177, 177L, Math 175, 176, SSH elective (2 cr.)
Industrial Engineering—I E 101, Psych 101, Sp Cm 212
Mechanical Engineering—Chem 167L, SSH elective (3 cr.)
Metallurgical Engineering—Chem 177*, 177L, free elective (2 cr.)
Nuclear Engineering—Nuc E 191, 192, Econ 201

The student's adviser may require or recommend courses in addition to those specified above if the preparation and progress of the student are such that additional courses are necessary or desirable.

Requirement for Entry into Professional Program
Students enrolled in the College of Engineering must satisfy both of the following requirements before being admitted to a professional program:

1. Completion of the basic program with an average of 2.00 or better in the basic program.

2. A cumulative grade average of 2.00 or better for all courses taken at Iowa State up to that time.

In some engineering programs, the number of students applying for admission to the professional program exceeds the number of students that can be accommodated. The limitation is determined by resources—faculty and laboratory facilities. The college has developed an enrollment management plan that is designed to allocate the available spaces on a basis that combines grades in the basic program courses with the amount of time the students have spent awaiting acceptance into the professional program. The details are available from the College Classification Office and will be explained thoroughly during an orientation program.

Engineering undergraduates must be admitted to a professional program before they may enroll in 200-level or above courses offered in the College of Engineering. The only exceptions to the application of this rule are the following:

a. Students who have completed all of their course work while enrolled in the College of Engineering, but have not been admitted to a professional program, may enroll for not more than one semester in 200-level or above courses offered by departments in the College of Engineering which have not been designated for enrollment management. This exception may be extended to two semesters for students whose curriculum requires Chem 178 and 178L.

b. Students transferring to the College of Engineering from another college or university, or from a program outside this college, who do not qualify for admission to a professional program may enroll for not more than two semesters in 200-level or above courses offered by departments in the College of Engineering which have not been designated for enrollment management.

c. Iowa State students not pursuing an engineering degree may take engineering courses without restrictions provided they meet the prerequisites and space is available.

d. Only the first two semesters of 200-level and above engineering courses, taken at ISU while a student is not enrolled in the College of Engineering, can be applied toward an engineering degree.

Students reentering the College of Engineering must have the approval of the College Academic Standards Committee.

Requirement for Graduation
In order to graduate in a professional engineering curriculum, a student must have a minimum GPA of 2.00 in a department-designated group of 200-level and above courses. These courses will total not less than 24 nor more than 48 semester credits.

Advising System
The purpose of the advising system in the College of Engineering is to work constructively with students in developing their individual academic programs and to maintain close contact with students during their college careers.

The college offers an orientation program during the spring and summer for students planning to enter in the fall and during the fall for students planning to enter in the spring. All entering students are encouraged to attend an orientation session. Tests given at this time help determine the student's level of achievement and enable the adviser to prepare an appropriate program for the student.

Special Programs
Engineering College students may participate in the following undergraduate programs. These programs are integrated into the professional engineering curricula and often require additional work. Each individual program is developed by the student and her/his engineering adviser.

a. Cooperative Education Program—The College of Engineering offers, through its curricula, cooperative programs in which students may gain practical experience in engineering during college years.

These programs are arranged so that the academic work is taught at the university and practical experience is gained by working in industry during certain periods each year. The student under a cooperative program receives experience in a chosen profession, plus financial return.

The employer can evaluate the student's potential as a possible future permanent employee. The college gains by the engineering experiences which the cooperative student brings into the classroom.

In general, students under these programs will require one year more to complete the usual curriculum requirements. The first contact with industry usually comes after completion of the first or second year. The college does not guarantee the kind of work or wages, but attempts to place students to their best educational and financial advantages.

A student must observe regulations of the employer and must not expect special treatment. University holidays do not apply to cooperative students, nor are students allowed time off for university activities. A student may not enroll in classes at any educational institution during a period of cooperative employment without university approval.

Those in the cooperative program are considered by the university to be students while they are employed. Such students are subject to university regulations concerning conduct during this period and are liable to dismissal from the university for misconduct on the job. They may continue living in university housing during work periods.

Cooperative students pay no fees to the university during work periods but may attend student activities provided they pay the activity fee.

b. Engineering Journalism Program. See *Index*, also see *Engineering Operations*.

c. Environmental Studies Program. See *Index*.

d. Honors Program. The College of Engineering participates in the University Honors Program (see *Index*). In summary, the Honors Program is designed for students with above average ability who wish to individualize their programs of study. For further details consult the chair of the Engineering College Honors Program Committee or your departmental Honors Program adviser.

e. International Studies Program. See *Index*.

f. Officer Education Program (ROTC). See *Index*, also see *Engineering Operations*.

Curriculum in Aerospace Engineering

Leading to the degree bachelor of science. Total credits required: 135. See also *Basic Program* and *Cooperative Programs*.

Professional Program

Sophomore Year

Cr.	Fall
4	Elementary Multivariable Calculus—Math 265
5	Introduction to Classical Physics II—Phys 222
3	Statics of Engineering—E M 274
3	Aerodynamics I—Aer E 241*
3	Introduction to Aerospace Engineering—Aer E 202*
18	

Cr.	Spring
4	Elementary Differential Equations and Laplace Transforms—Math 267
3	Mechanics of Materials—E M 324
3	Dynamics—E M 345
3	Aerodynamics II—Aer E 242*
1	Aerospace Laboratory I—Aer E 271
3	SSH elective[1]
17	

Junior Year

Cr.	Fall
3	Aerodynamic Theory I—Aer E 341*
1	Aerodynamics Laboratory—Aer E 341L
3	Astrodynamics I—Aer E 351*
3	Flight Structures Analysis—Aer E 321*
1	Structures Laboratory—Aer E 321L
4	Engineering Thermodynamics I—M E 331
3	Materials for Aerospace Applications—M S E 371
18	

Cr. Spring
3 Aerodynamic Theory II—Aer E 342*
3 Flight Vehicle Stability and Control—Aer E 355*
R Flight Experience—Aer E 301
0.5 Gas Dynamics Laboratory—Aer E 342L
3 Advanced Flight Structures—Aer E 421*
3 Analytical Techniques for Aerospace Design—Aer E 361*
3 SSH elective[1]
15.5

Senior Year

Cr. Fall
3 Aerospace Vehicle Propulsion I—Aer E 411*
3 Technical elective[2]
3 Flight Control Systems I—Aer E 431*
3 Design and Analysis I—Aer E 461*
R Aerospace Seminar—Aer E 491
4 SSH elective[1]
16

Cr. Spring
3 Design and Analysis II—Aer E 462
R Aerospace Seminar—Aer E 492
8 Technical electives[2]
4 Introduction to Circuits, Instruments, and Electronics—E E 441
15

[1]The social sciences and humanities (SSH) electives are to be selected from the department-approved list of courses. Not to be taken under the P-NP policy.

[2]Technical electives are of four types: (1) mathematics, 3 credits; (2) physics, 3 credits; (3) aerospace, 3 credits; and (4) senior elective, 2 credits. At least one course must be chosen from department-approved lists for each type. These courses are not to be taken under the P-NP policy.

*Core professional curriculum. A student must have a minimum grade-point average of 2.00 in this group of courses in order to graduate.

Curriculum in Agricultural Engineering

With options in process engineering, agricultural power and machinery, structures and environment, soil and water control, and food engineering. Administered jointly by the College of Agriculture and the College of Engineering. Leading to the degree bachelor of science. Total credits required: 134.5. See also *Basic Program* and *Cooperative Programs*.

Sophomore Year

Cr. Fall
2 or 3 Agricultural Engineering Laboratory I—A E 213 or Principles of Biology—Biol 110[3]
3 Statics of Engineering—E M 274*
4 Elementary Multivariable Calculus—Math 265*
5 Introduction to Classical Physics II—Phys 222*
3 SSH elective[1]
17 or 18

Cr. Spring
3 Computer Applications in Agricultural Engineering—A E 202*
2 Agricultural Engineering Laboratory II—A E 214*
4 Elementary Differential Equations and Laplace Transforms—Math 267*
3 Mechanics of Materials—E M 324*
3 Principles of Macroeconomics—Econ 205
3 Fundamentals of Botany—Bot 307 (omit for food engineering option)
15 or 18

Junior Year

Cr. Fall
3 Modeling Agricultural Engineering Systems—A E 301*
3 Agri-Industrial Applications of Electric Power—A E 363
1 Mechanics of Materials Lab—E M 327*
3 Thermodynamics—M E 330*
3 SSH elective[1]
4 Option requirement[2]
17

Cr. Spring
1 Seminar—A E 302
9 or 11[3] Option requirement[2]
3 Mechanics of Fluids—E M 378* or Momentum and Transport Operations—Ch E 356*[3]
3 Fundamentals of Speech Communication—Sp Cm 211
16 or 18

Senior Year

Cr. Fall
13 Option requirement[2]
R Seminar—A E 401
3 SSH elective[1]
16

Cr. Spring
4 Agricultural Engineering Design—A E 446
9 Option requirement[2]
4 SSH elective[1]
17

[1]Social sciences and humanities (SSH) sequences are to be chosen from the department-approved list.

[2]In the junior and senior years, each student elects one of the options and takes the courses listed for the selected option. The elective courses must be selected in consultation with the student's adviser to ensure that ABET requirements for engineering science and engineering design are met.

[3]Food engineering option only.
*Core professional curriculum. A student must have a minimum grade-point average of 2.00 in this group of courses in order to graduate.

Options

Process Engineering—A E 342, 359L, 421, 478, 495; Cpr E 340, 440; E M 345; M E 336, 441; and 6 credits from A E 503, 569, 571, 572; C E 326; Econ 335.

Agricultural Power and Machinery—A E 342, 359L, 421, 447, 478; Cpr E 340 or E E 441; EM 345; M E 310, 321, 416; M S E 270 and 4 credits from A E 503, 569; E M 417, 425, 444; M E 322, 336.

Structures and Environment—A E 342, 359L, 421, 478; C E 332, 334; Cpr E 340 or E E 441; E M 345; M E 336, 441, and 6 credits from A E 571, 572; C E 333; E M 425; M E 442.

Soil and Water Control—A E 342, 421, 478; C E 211, 332, 334, 360, 372; and 10 credits from A E 359L, 522; Agron 318, 354; C E 326, 528; Cpr E 340.

Food Engineering—A E 495, 496; B B 221; Ch E 357; Cpr E 340; E M 345; F Tch 201, 311, 421; Micro 330; and 8 credits from A E 421, 478, 569; C E 326; Ch E 358; M E 441; Cpr E 440.

Curriculum in Ceramic Engineering

Administered by the Department of Materials Science and Engineering. Leading to the degree bachelor of science. Total credits required: 132.5. See also *Basic Program*.

Professional Program

Sophomore Year

Cr. Fall
4 Introduction to Ceramic Engineering—M S E 230*
1 Ceramic Processing Laboratory—M S E 230L*
5 Introduction to Classical Physics II—Phys 222
4 Elementary Multivariable Calculus—Math 265
R Seminar—M S E 210
3 General Chemistry—Chem 178
1 Laboratory in General Chemistry—Chem 178L
18

Cr. Spring
4 Introduction to Ceramic Science—M S E 231*
3 Elementary Differential Equations—Math 266
4 Priniciples of Economics—Econ 201
4 Instruments for Materials Characterization—M S E 244*
R Seminar—M S E 211
15

Junior Year

Cr. Fall
3 High Temperature Technology—M S E 330*
3 Thermochemistry for Materials Science and Engineering—M S E 360*
4 Electricity and electronics elective[2]
3 Statics of Engineering—E M 274
4 Electronic Ceramics—M S E 343*
R Seminar—M S E 310
17

Cr. Spring
3 Engineering Statistics—Stat 305
3 High Temperature Processes—M S E 345*
1 High Temperature Processing Laboratory—M S E 345L*
3 Vitreous State—M S E 342*
1 Vitreous State Laboratory—M S E 342L*
3 Mechanics of Materials—E M 324
R Inspection Trip—M S E 340
1 Mechanics of Materials Laboratory—E M 327
R Seminar—M S E 311
15

Senior Year

Cr. Fall
3 Mechanical and Thermal Properties of Ceramic Materials—M S E 440*
3 Technical Communications—Engl 314
3 Ceramic Engineering Design—M S E 445*
3 Technical elective[3]
2 Analysis for Engineering Economy—I E 304
3 SSH elective[1]
R Seminar—M S E 410
17

Cr. Spring
3 Refractories—M S E 441*
3 Ceramic Engineering Design—M S E 446*
3 Technical elective[3]
6 SSH electives[1]
2 Application of Statistics to Materials—M S E 341*
R Seminar—M S E 411
17

[1]Social sciences and humanities (SSH) electives must be department approved.

[2]E E 441 or Cpr E 340.

[3]Technical electives must be department approved.

*Core professional curriculum. A student must have a minimum grade-point average of 2.00 in this group of courses in order to graduate.

Curriculum in Chemical Engineering

Leading to the degree bachelor of science. Total credits required: 133.5. See also *Basic Program* and *Cooperative Programs.*

Professional Program

Sophomore Year

Cr. Fall
3 Material and Energy Balances—Ch E 210*
4 Elementary Multivariable Calculus—Math 265
5 Introduction to Classical Physics I—Phys 221
4 SSH elective[1]
16

Cr. Spring
3 Momentum Transport Operations—Ch E 356*
2 Design Analysis Laboratory—Ch E 230*
4 Fundamentals of Mechanics—E M 301
4 Elementary Differential Equations and Laplace Transforms—Math 267
5 Introduction to Classical Physics II—Phys 222
18

Junior Year

Cr. Fall
3 Heat and Mass Transfer—Ch E 357*
4 Chemical Engineering Thermodynamics—Ch E 381*
1 Chemical Engineering Laboratory I—Ch E 324*
3 Physical Chemistry—Chem 321
3 Organic Chemistry—Chem 331
3 SSH elective[1]
17

Cr. Spring
R Seminar—Ch E 302
3 Mass Transfer Operations—Ch E 358*
1 Chemical Engineering Laboratory II—Ch E 325*
3 Chemical Reactor Design—Ch E 382*
3 Organic Chemistry—Chem 332
3 Physical Chemistry—Chem 322
2 Laboratory in Physical Chemistry—Chem 321L
3 SSH elective[1]
18

Senior Year

Cr. Fall
R Seminar—Ch E 401
1 Chemical Engineering Laboratory III—Ch E 426*
3 Process Controls—Ch E 421*
6 Professional electives[2]
3 Technical Communications—Engl 314
3 SSH elective[1]
16

Cr. Spring
3 Process and Plant Design—Ch E 430*
6 Professional electives[2]
4 Introduction to Circuits, Instruments and Electronics—E E 441
3 SSH elective[1]
16

[1]Selected from list of department-approved social sciences and humanities (SSH) courses.

[2]Selected to develop a professional emphasis subject to departmental restrictions.

*Core professional curriculum. A student must have a minimum grade-point average of 2.00 in this group of courses in order to graduate.

Curriculum in Civil Engineering

Administered by the Department of Civil and Construction Engineering

Leading to the degree bachelor of science. Total credits required: 133.5. See also *Basic Program* and *Cooperative Programs.*

For those interested in construction engineering a curriculum is provided which leads to the degree bachelor of science in construction engineering. For particulars, see *Curriculum in Construction Engineering.*

Sophomore Year

Cr. Fall
4 Elementary Multivariable Calculus—Math 265
5 Introduction to Classical Physics II—Phys 222
3 Fundamentals of Surveying and Computer Applications I—C E 211
1 The Practice of Engineering in Government—C E 295[2]
5 Statics and Dynamics—E M 307*
17-18

Cr. Spring
3 Elementary Differential Equations—Math 266
3 Fundamentals of Surveying and Computer Applications II—C E 214*
1 The Private Practice of Engineering—C E 296[2]
3 Mechanics of Materials—E M 324*
1 Mechanics of Materials Laboratory—E M 327
2 Geology for Engineers—Geol 301
3 Elective[1]
15-16

Junior Year

Cr. Fall
3 Water Quality Engineering—C E 326
3 Structural Analysis I—C E 332*
3 Mechanics of Fluids—E M 378*
3 Soil Engineering—C E 360*
2 Analysis for Engineering Economy—I E 304
3 Elective[1]
17

Cr. Spring
3 Structural Steel Design I—C E 333
3 Introduction to Transportation Engineering—C E 351*
3 Design of Concretes and Pavement Structures—C E 382
4 Engineering Hydrology and Hydraulics—C E 372*
4 Elective[1]
17

Senior Year

Cr.	Fall
3	Reinforced Concrete Design I—C E 334
3	Highway Design—C E 452
3	Engineering Construction—C E 485
8	Electives[1]
17	

Cr.	Spring
3	Construction or management elective[1]
14	Electives[1]
17	

[1]Shall be chosen from department and curriculum-approved lists. Electives shall include: (1) 17 credits of social sciences or humanities studies; (2) 6 credits of engineering science; (3) 6 credits of technical elective, including at least one course taken in the Department of Civil and Construction Engineering; (4) 3 credits of communications elective. The construction or management electives must be selected in construction engineering, economics, psychology, business, and/or industrial engineering. Students appointed to advanced ROTC may substitute 4 credits of advanced ROTC for 4 credits of technical electives.

[2]C E 295 or 296 required, but not both.

*Core professional curriculum. A student must have a minimum grade-point average of 2.00 in this group of courses in order to graduate.

Curriculum in Computer Engineering

Administered by the Department of Electrical Engineering and Computer Engineering.

Leading to the degree bachelor of science. Total credits required: 137.5. See also *Basic Program* and *Cooperative Programs*.

Sophomore Year

Cr.	First Semester
3	Introduction to Digital Techniques—Cpr E 280*
3	Electric Circuits I—E E 205*
2	Electrical Instrumentation and Experimentation—E E 235*
4	Elementary Multivariable Calculus—Math 265, or Linear Algebra, Multivariable Calculus and Differential Equations—Math 270[4]
5	Introduction to Classical Physics II—Phys 222
R	Professional Programs Orientation—E E 261
17	

Cr.	Second Semester
3	Electric Circuits II—E E 206*
1	Digital Laboratory I—Cpr E 287*
4	Introduction to Microprocessors—Cpr E 288*
4	Elementary Differential Equations and Laplace Transforms—Math 267, or Linear Algebra, Multivariable Calculus and Differential Equations—Math 371[4]
3	Fundamentals of UNIX and C Language—Cpr E 284*
3	SSH elective[1]
18	

Junior Year

Cr.	First Semester
3	Computer Organization and Design I—Cpr E 384*
4	Electronics I—E E 330*
3	Elementary Electromagnetics I—E E 212*
4	Technical Elective[2]
3	SSH elective[1]
17	

Cr.	Second Semester
4	Introduction to the Design of Computer-Based Systems—Cpr E 389*
3	Probability and Statistics for Electrical and Computer Engineers—Stat 333
4	Digital Integrated Circuits—E E 436*
3	Computer Organization and Design II—Cpr E 385*
3	SSH elective[1]
17	

Senior Year

Cr.	First Semester
3	Digital Systems Design Laboratory I—Cpr E 481
5	Technical elective[2]
3	Engineering science elective[3]
3	Mathematics elective[4]
3	SSH elective[1]
17	

Cr.	Second Semester
3	Digital Systems Design Laboratory II—Cpr E 482
7	Technical electives[2]
6	SSH electives[1]
16	

[1]Social sciences and humanities (SSH) electives must be chosen from a list of courses and sequences of courses approved by the department; pass/not pass credit not accepted.

[2]Technical electives must be chosen to satisfy departmental requirements concerning content, distribution, level, and the engineering science and engineering design requirements. All technical electives must be chosen from a list approved by the department. At least 6 credits must be selected from computer science. Details are available in the departmental Undergraduate Advising Center. Pass/not pass credit not accepted. Sixteen credits of technical electives are required. This number may be increased by one credit depending on the mathematics elective (see below).

[3]Engineering science elective must be chosen from a list approved by the department; pass/not pass credit not accepted. This elective must be from another engineering department.

[4]If a student has completed the Math 175, 176, 270 and 371 series, then the mathematics elective requirement is satisfied by the two extra credits of Math 175, 176 and by one additional technical elective credit. Otherwise a student must choose one of the following math courses (pass/not pass credit not accepted): Math 307, 471, or 481.

*Core professional curriculum. A student must have a minimum grade-point average of 2.00 in this group of courses in order to graduate.

Curriculum in Construction Engineering

Administered by the Department of Civil and Construction Engineering.

Leading to the degree bachelor of science. Total credits required: 135.5 Building emphasis; 136.5 Heavy emphasis; 134.5 Mechanical emphasis. See also *Basic Program* and *Cooperative Programs*.

Sophomore Year

Cr.	Fall
3	Fundamentals of Surveying and Computer Applications I—C E 211
3	Construction Materials and Methods—Con E 241
4	Elementary Multivariable Calculus—Math 265
5	Introduction to Classical Physics II—Phys 222
3	Fundamentals of Speech Communication—Sp Cm 211
18	B,H,M

Cr.	Spring
3	Fundamentals of Surveying and Computer Applications II—C E 214 (H)
3	Construction Contract Documents—Con E 245
4	Principles of Economics—Econ 201
3	Statics of Engineering—E M 274*
2	Geology for Engineers—Geol 301 (B,H)
3	Elementary Differential Equations Math 266
3	Financial Accounting—Acct 284 (B, M)
18	B; 16 M; 18 H

Junior Year

Cr.	Fall
3	Soil Engineering—C E 360* (B,H)
3	Construction Estimating—Con E 346*
3	Mechanics of Materials—E M 324*
3	Mechanics of Fluids—E M 378*
2	Analysis for Engineering Economy—I E 304
4	Engineering Thermodynamics I—M E 331* (M)
2	Introduction to Statistics—Stat 105
16	B,H; 17 M

Cr.	Spring
3	Structural Analysis I—C E 332*
3	Contractor Organization and Management of Construction—Con E 371
3	Construction Equipment and Heavy Construction Methods—Con E 372* (B,H)
3	Engineering Materials—E M 337
3	Elements of Heat Transfer—M E 336* or Heat Transfer—M E 436* (M)
3	Business Communication—Engl 302[3]
3	SSH elective[2]
18	B,H,M

Senior Year

Cr. Fall
3 Structural Steel Design I—C E 333 (B,H)
3 Financial Accounting—Acct 284 (H)
1 Professional Development—Con E 410
3 Concrete and Steel Construction—Con E 440* (B,H)
4 Introduction to Circuits, Instruments, and Electronics—E E 441 (B,M)
3 Engineering Law—I E 480
3 Fundamentals of Heating, Ventilating, and Air Conditioning—M E 441* (M)
3 Engineering design elective (M)[1]
3 SSH elective[2]
17 B,M; 16 H

Cr. Spring
3 Reinforced Concrete Design I—C E 334 (B,H)
3 Design of Concretes and Pavement Structures—C E 382 (H)
3 Foundations—C E 460 (H)
3 Construction Planning, Scheduling and Control—Con E 441
2 Introduction to Electric Machinery—E E 447 (M)
3 SSH elective[2]
4 Principles of Heating and Air Conditioning—M E 440 (B)
3 Heating and Air Conditioning Design—M E 442 (M)
3 Engineering design elective (M)[1]
2 Basic science elective (M)[1]
3 Business management elective[1] (B,H)
16 B; 18 H; 16 M

B - Building construction emphasis.
H - Heavy construction emphasis.
M - Mechanical construction emphasis.

[1]Chosen from curriculum-approved lists. All electives must be taken for a grade. Pass-Not Pass grades are not acceptable.

[2]Social sciences and humanities (SSH) electives chosen from curriculum-approved list. One of these must have a prerequisite of Psych 101, Econ 201, or a previously taken social sciences and humanities elective.

[3]All English courses taken, including those in the Basic Program, require a C or better. C− or less requires additional composition coursework. All electives must be taken for a grade. Pass/Not Pass grades are not acceptable.

*Core professional curriculum. A student must have a minimum grade-point average of 2.00 in this group of courses in order to graduate.

Undesignated courses are for all emphases.

Curriculum in Electrical Engineering

Administered by the Department of Electrical Engineering and Computer Engineering. Leading to the degree bachelor of science. Total credits required: 137.5. See also *Basic Program* and *Cooperative Programs*.

Sophomore Year

Cr. **First Semester**
3 Electric Circuits I—E E 205*
2 Electrical Instrumentation and Experimentation—E E 235*
3 Introduction to Digital Techniques—Cpr E 280*
4 Elementary Multivariable Calculus—Math 265, or Linear Algebra, Multivariable Calculus and Differential Equations—Math 270
5 Introduction to Classical Physics II—Phys 222
R Professional Programs Orientation—E E 261
17

Cr. **Second Semester**
3 Electric Circuits II—E E 206*
3 Elementary Electromagnetics I—E E 212*
4 Electronics I—E E 330*
4 Elementary Differential Equations and Laplace Transforms—Math 267, or Linear Algebra, Multivariable Calculus and Differential Equations—Math 371
3 SSH elective[1]
17

Junior Year

Cr. First Semester
4 Elementary Electromagnetics II—E E 313*
4 Electronics II—E E 331*
3 Linear Systems, Continuous-Time and Discrete-Time—E E 374*
4 Intermediate Engineering Mathematics—Math 395
3 SSH elective[1]
18

Cr. Second Semester
4 Electromagnetic Devices and Electric Machinery—E E 351*
3 Electric Network Design—E E 309*
4 Elementary Modern Physics—Phys 324
3 Probability and Statistics for Electrical and Computer Engineers—Stat 333
3 Introduction to Scientific Computation—Math 473
17

Senior Year

Cr. First Semester
3 Electrical Systems Design I—E E 461
7 Technical electives[2]
3 Engineering science elective[3]
3 SSH elective[1]
16

Cr. Second Semester
3 Electrical Systems Design II—E E 462
7 Technical electives[2]
3 SSH elective[1]
3 SSH elective[1]
16

[1]Social sciences and humanities (SSH) electives must be chosen from a list of courses and sequences of courses approved by the department; pass/not pass credit not accepted.

[2]Technical electives are of two types: (1) courses in computer engineering and electrical engineering, and (2) other courses in engineering and science. All technical electives must be chosen from lists approved by the department. Technical electives must be chosen to satisfy departmental requirements concerning content, distribution, level, and the engineering science and engineering design requirements. Details are available in the departmental Undergraduate Advising Center. Pass/not pass credit not accepted.

[3]Engineering science elective must be chosen from a list approved by the department; pass/not pass credit not accepted. This elective must be from another engineering department.

*Core professional curriculum. A student must have a minimum grade-point average of 2.00 in this group of courses in order to graduate.

Curriculum in Engineering Operations

Administered by the Department of Industrial Engineering.

Leading to the degree bachelor of science. Total credits required: 121.5. See also *Cooperative Program*.

In this era of rapid technological change, there is an expanding and continually accelerating need for persons with an engineering background. Engineering operations is specifically designed to develop this background within several engineering disciplines, or in combination with engineering and other disciplines.

The curriculum consists of a basic core of required courses in the sciences, engineering, and management to which are added 62 credits of elective courses in the specific categories of engineering, social sciences and humanities (SSH), vocational, and preliminary supporting subjects. Within this framework, students may specialize toward specific occupational objectives of their choice. In the course of achieving individual student's objectives, the total credits may well exceed the minimum requirements of 121.5.

Prior to entering the engineering operations program the student must have completed the basic program and have presented a description of the vocational objective to be achieved through the program to the department chairman for approval. In addition the student will submit a schedule of courses to support this objective. Until admitted to the program, students will be considered to be pre-engineering or in a pre-professional or professional program of another curriculum.

Special Programs

To meet special needs, a special program is available in engineering journalism (see *Index*).

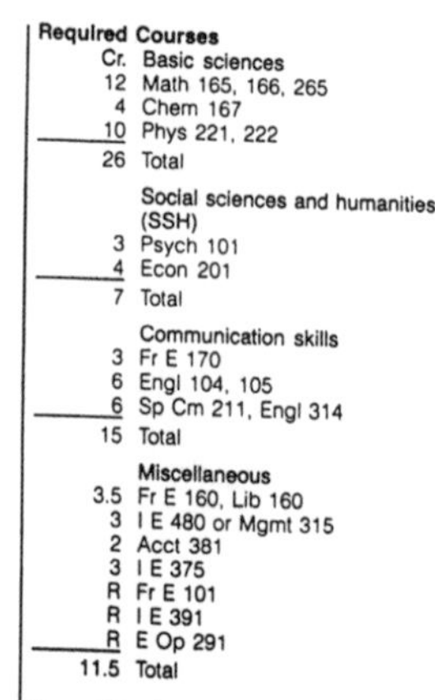

Required Courses

Cr.	Basic sciences
12	Math 165, 166, 265
4	Chem 167
10	Phys 221, 222
26	Total
	Social sciences and humanities (SSH)
3	Psych 101
4	Econ 201
7	Total
	Communication skills
3	Fr E 170
6	Engl 104, 105
6	Sp Cm 211, Engl 314
15	Total
	Miscellaneous
3.5	Fr E 160, Lib 160
3	I E 480 or Mgmt 315
2	Acct 381
3	I E 375
R	Fr E 101
R	I E 391
R	E Op 291
11.5	Total

Group Requirements
Course combinations for each student should be integrated toward a vocational objective. Each student's choice of courses in the following groups must be approved in advance by the head of the Department of Industrial Engineering.

Cr.	
12	**Engineering science:** engineering mechanics, 6; electrical engineering, 4; measurements, 2
18	**Engineering analysis** (300 level or above)
12	**Supporting** (basic and engineering sciences)
10	**Vocational** (300 level or above)
10	**Social sciences and humanities (SSH) sequences**

Curriculum in Engineering Science

Administered by the Department of Engineering Science and Mechanics. Leading to the degree bachelor of science. Total credits required: 138.5. See also *Basic Program* and *Cooperative Programs*.

Sophomore Year

Cr.	Fall
4	Linear Algebra, Multivariable Calculus and Differential Equations—Math 270*
5	Introduction to Classical Physics II—Phys 222
4	Introduction to Mechanics I—E M 201*
3	Electric Circuits I—E E 205*
2	Electrical Instrumentation and Experimentation—E E 235*
R	Engineering Science Seminar—E Sci 410
18	

Cr.	Spring
4	Linear Algebra, Multivariable Calculus and Differential Equations—Math 371*
3	Physical science elective[1]
4	Introduction to Mechanics II—E M 202*
3	Communications skills elective[2]
4	Electronics I—E E 330*
R	Engineering Science Seminar—E Sci 410
18	

Junior Year

Cr.	Fall
4	Probability and Statistical Inference for Engineers—Stat 231
4	Engineering Materials I—E Sci 351*
3	Numerical Methods in Engineering Science and Mechanics—E Sci 381*
3	Intermediate Solid Mechanics—E M 401*
3	SSH elective[3]
R	Engineering Science Seminar—E Sci 410
17	

Cr.	Spring
3	Introduction to Partial Differential Equations—Math 385*
3	Experimental Methods in Engineering Science and Mechanics—E Sci 382
3	Depth elective[4]
4	SSH elective[3]
4	Engineering Thermodynamics I—M E 331*
R	Engineering Science Seminar—E Sci 410
17	

Senior Year

Cr.	Fall
3	Intermediate Dynamics—E M 402*
3	Intermediate Fluid Mechanics—E M 403*
3	Heat Transfer—M E 436*
3	Technical elective[1]
2	Senior Engineering Science Design Project I—E Sci 481
3	SSH elective[3]
R	Engineering Science Seminar—E Sci 410
17	

Cr.	Spring
3	Mathematics elective[1]
3	Depth elective[4]
4	SSH elective[3]
3	Technical elective[1]
4	Senior Engineering Science Design Project II—E Sci 482
R	Engineering Science Seminar—E Sci 410
17	

[1]See department lists for approved mathematics electives, physical science electives, and technical electives.

[2]Any of the following courses are acceptable for satisfying the communications skills elective: Engl 204, 220, 302, 314, 415; Sp Cm 211, 212.

[3]These electives are to be chosen from the department-approved list of social sciences and humanities (SSH) courses. At least one sequence consisting of two or three courses is required as part of the 16 credits of social sciences and humanities electives. In addition, at least 6 credits of the required 16 must be in the humanities (art, English, foreign languages, history, philosophy, music, or religion).

[4]Each student must develop an area of specialization in one of the engineering sciences (solids, dynamics, fluids, material sciences, electrical science, or thermal sciences). This will be accomplished by taking two elective courses in the selected area of specialization. See department list for suggested lists of courses.

*Core professional curriculum. A student must have a minimum grade-point average of 2.00 in this group of courses in order to graduate.

Curriculum in Industrial Engineering

Leading to the degree bachelor of science. Total credits required: 134.5. See also *Basic Program* and *Cooperative Programs*.

Sophomore Year

Cr.	Fall
2	Industrial Accounting—Acct 381
4	Probability and Statistical Inference for Engineers—Stat 231*
4	Elementary Multivariable Calculus—Math 265
5	Introduction to Classical Physics II—Phys 222
3	Introduction to Industrial Engineering—I E 201*
18	

Cr.	Spring
3	Materials Science and Engineering—M S E 271
3	Applied Ergonomics—I E 277*
4	Principles of Economics—Econ 201
2	Industrial Computer Techniques—I E 239*
3	Elementary Differential Equations—Math 266*
R	Seminar—I E 293
15	

Junior Year

Cr.	Fall
3	Engineering Economy—I E 404*
3	Methods Engineering and Work Measurement—I E 373*
3	Statics of Engineering—E M 274
3	Linear Programming—I E 312*
4	Digital Circuits and Systems—Cpr E 340
R	Industrial Inspection Trip—I E 391
16	

Cr.	Spring
3	Industrial Methodology—I E 374
3	Quality Control—I E 361*
3	Stochastic Analysis—I E 313
4	E E elective[1]
3	Mechanics of Materials—E M 324
R	Professional Development—I E 392
16	

Senior Year

Cr.	Fall
6	Industrial Engineering Design—I E 441
3	Material and Project Control—I E 341
3	Thermodynamics—M E 330
3	SSH elective[2]
2	Human Resource Management—I E 424
17	

Cr.	Spring
6	Industrial engineering electives[1]
3	Technical elective[1]
6	SSH electives[2]
3	Technical Communications—Engl 314
18	

*Core professional curriculum. A student must have a minimum grade-point average of 2.00 in this group of courses in order to graduate.

[1]These electives are to be chosen from department-authorized lists with advance approval.

[2]These social sciences and humanities (SSH) electives must be chosen from a department-approved list and must include at least one 6-credit sequence of prerequisite or related courses. The sequence can include Psych 101 or Econ 201. No more than one 100-level course is to be included.

Curriculum in Mechanical Engineering

Leading to the degree bachelor of science. Total credits required: 133.5. See also *Basic Program* and *Cooperative Programs*.

Sophomore Year

Cr.	First Semester
4	Elementary Multivariable Calculus—Math 265
5	Introduction to Classical Physics II—Phys 222
3	Statics of Engineering—E M 274
4	SSH electives[1]
16	

Cr.	Second Semester
4	Elementary Differential Equations and Laplace Transforms—Math 267
3	Dynamics—E M 345*
3	Mechanics of Materials—E M 324*
2	Principles of Materials Science—M S E 270*
4	Mechanisms—M E 310*
16	

Junior Year

Cr.	First Semester
3	Engineering Statistics—Stat 305
3	Mechanical Behavior of Materials—M E 321*
4	Engineering Thermodynamics I—M E 331*
4	Introduction to Circuits, Instruments, and Electronics—E E 441*
3	SSH elective[1]
17	

Cr.	Second Semester
3	Mechanical Systems—M E 311*
3	Manufacturing Processes—M E 322*
3	Engineering Thermodynamics II—M E 332*
3	Fluid Flow—M E 335*
2	Introduction to Electric Machinery—E E 447
3	Technical Communications—Engl 314
R	Mechanical Engineering Seminar—M E 302
17	

Senior Year

Cr.	First Semester
3	Design of Machine Elements I—M E 416*
3	Engineering Measurements and Instrumentation—M E 360*
3	Heat Transfer—M E 436*
3	SSH elective[1]
5	Technical electives[2]
17	

Cr.	Second Semester
3	Design elective[3]
2	Experimental Engineering—M E 460
9	Technical electives[2]
3	SSH elective[1]
17	

[1]Social sciences and humanities (SSH) elective courses must be chosen from a department-approved list and must include either Econ 201, 205, or 206, at least one 6-credit sequence; at least 6 credits in the humanities and at least 6 credits in the social sciences.

[2]Technical electives are to include at least 9 credits of mechanical engineering courses from 400- and 500-level offerings. All technical electives must be chosen from a department-approved list. Suggested areas of specialization are the following:

Energy conversion and utilization—M E 444, 446, 447, 448, 449; Acct 381; E E 456, 457; I E 404.
Machines and systems—M E 411, 412, 414, 415, 417, 420, 470, 490F, 514, 515, 516, 518, 535; E M 514, 515, 517, 519, 525, 544, 584.
Materials and Manufacturing—M E 411, 490G, 515, 520, 521, 526; E M 514, 544; M S E 401, 402, 522, 524; I E 475.
Thermal and environmental engineering—M E 441, 442, 444, 445, 446, 447, 475, 490D, 490J, 490K, 530, 531, 532, 533, 536, 538, 540, 541, 542, 546, 547, 548, and applicable courses in other departments.
Propulsion—M E 445, 447, 448, 449, 490J, 490K, 490L, 542, 545, 548; Aer E 411.

[3]The design elective must be chosen from M E 415, 442, 446, or 449.

*Core professional curriculum. A student must have a minimum grade-point average of 2.00 in this group of courses in order to graduate.

Curriculum in Metallurgical Engineering

Administered by the Department of Materials Science and Engineering. Leading to the degree bachelor of science. Total credits required: 129.5. See also *Basic Program*.

Professional Program

Sophomore Year

Cr.	Fall
2	Principles of Materials Science—M S E 270*
2	Introductory Physical Metallurgy Lab—M S E 270L*
5	Introduction to Classical Physics II—Phys 222*
3	General Chemistry—Chem 178
1	Laboratory in General Chemistry—Chem 178L
4	Elementary Multivariable Calculus—Math 265
17	

Cr.	Spring
3	Introduction to Materials Processing—M S E 203*
3	Statics of Engineering—E M 274*
3	Elementary Differential Equations—Math 266
4	SSH elective[1]
3	Engineering Statistics—Stat 305
16	

Junior Year

Cr.	Fall
3	Thermochemistry for Materials Science and Engineering—M S E 360*
4	Physical Metallurgy—M S E 301*
3	Physical Metallurgy Laboratory I—M S E 301L*
3	Mechanics of Materials—E M 324
3	SSH elective[1]
16	

Cr.	Spring
4	Principles of Extractive Metallurgy—M S E 361*
4	Physical Metallurgy—M S E 302*
3	Physical Metallurgy Laboratory II—M S E 302L*
3	Mechanical Metallurgy—M S E 401
3	SSH elective[1]
17	

[1]Selected from a list of department-approved courses.

[2]Technical electives must include 3 credits of advanced basic science to meet ABET requirements.

*Core professional curriculum. A student must have a minimum grade-point average of 2.00 in this group of courses in order to graduate.

Senior Year

Cr. Fall
- 3 Mechanical Metallurgy—M S E 402*
- 3 English composition elective
- 3 SSH elective[1]
- 2 Mechanical Metallurgy Laboratory—M S E 402L*
- 4 Introduction to Circuits, Instruments, and Electronics—E E 441

15

Cr. Spring
- 3 Metallurgical Engineering Design—M S E 421
- 3 SSH elective[1]
- 8 Technical electives[2]
- 2 Free elective

16

[1]These electives are to be chosen from the department-approved list of social sciences and humanities (SSH) courses.

[2]These electives are to be chosen from the department-approved list of technical courses.

*Core professional curriculum. A student must have a minimum grade-point average of 2.00 in this group of courses in order to graduate.

Curriculum in Nuclear Engineering

Leading to the degree bachelor of science. Total credits required: 134.5. See also *Basic Program* and *Cooperative Programs*.

Sophomore Year

Cr. Fall
- 4 Elementary Multivariable Calculus—Math 265
- 5 Introduction to Classical Physics II—Phys 222
- 2 Introductory Laboratory in Nuclear Engineering—Nuc E 261*
- R Sophomore Seminar I—Nuc E 291
- 3 Statics of Engineering—EM 274
- 2 Introduction to Nuclear Engineering Computations—Nuc E 201

16

Cr. Spring
- 4 Elementary Differential Equations and Laplace Transforms—Math 267
- 3 Fundamentals of Nuclear Engineering—Nuc E 211*
- R Sophomore Seminar II—Nuc E 292
- 3 Materials Science and Engineering—M S E 271
- 3 Mechanics of Fluids—EM 378
- 3 Mechanics of Materials—EM 324

16

Junior Year

Cr. Fall
- 3 Fission Reactor Analysis I—Nuc E 331*
- 3 Radiation Detection and Measurement—Nuc E 361*
- R Junior Seminar—Nuc E 391
- 3 Nuclear Materials and Radiation Effects—Nuc E 375*
- 3 Thermodynamics—M E 330
- 2 Fundamentals of Public Speaking—Sp Cm 212
- 3 SSH elective[1]

17

Cr. Spring
- 3 Fission Reactor Analysis II—Nuc E 332*
- 3 Elements of Heat Transfer—ME 336
- 4 Introduction to Circuits, Instruments, and Electronics—E E 441
- 2 Laboratory in Reactor Analysis—Nuc E 362*
- R Junior Seminar and Inspection Trip—Nuc E 392
- 2 Analysis for Engineering Economy—I E 304
- 3 SSH elective[1]

17

Senior Year

Cr. Fall
- 3 Nuclear Thermal-Hydraulic Analysis and Design—Nuc E 481*
- R Senior Seminar I—Nuc E 491
- 3 Technical Communications—Engl 314
- 3 Nuclear Fuel Cycles, Processes, and Management—Nuc E 451*
- 2 Nuclear Reactor Design—Nuc E 482*
- 1 Nuclear Systems Laboratory—Nuc E 461*
- 3 Technical elective[2]
- 3 SSH elective[1]

18

Cr. Spring
- 3 Safety and Control of Nuclear Systems—Nuc E 441*
- 3 Nuclear Engineering Design—Nuc E 485*
- 2 Nuclear Reactor Design—Nuc E 483*
- R Senior Seminar—Nuc E 492
- 6 Technical electives[2]
- 3 SSH elective[1]

17

[1]These electives are to be chosen from the department-approved list of social sciences and humanities (SSH) courses.

[2]These electives are to be chosen from the department-approved list of technical courses.

*Core professional curriculum. A student must have a minimum grade-point average of 2.00 in this group of courses in order to graduate.

APPENDIX D

Engineering Societies

The Junior Engineering Technical Society (JETS), 1420 King Street, Alexandria, VA 22314-2715, supplies information and guidance from all the societies in the following list. Requests for brochures should be accompanied by a stamped self-addressed envelope.

American Academy of Environmental Engineers
132 Holiday Court, Suite 206
Annapolis, MD 21401

American Ceramic Society
757 Brooksedge Plaza Drive
Westerville, OH 43081

American Congress on Surveying and Mapping
5410 Grosvenor Lane
Suite 100
Bethesda, MD 20814-2122

American Institute of Aeronautics and Astronautics, Inc.
370 L'Enfant Promenade SW
Washington, DC 20024

American Institute of Chemical Engineers
345 East 47th Street
New York, NY 10017

American Institute of Mining, Metallurgical and Petroleum Engineers
345 East 47th Street
New York, NY 10017

American Nuclear Society
555 North Kensington Avenue
LaGrange Park, IL 60525

American Society for Engineering Education
11 Dupont Circle, Suite 200
Washington, DC 20036

The American Society of Agricultural Engineers
2950 Niles Road
St. Joseph, MI 49085

American Society of Civil Engineers
345 East 47th Street
New York, NY 10017

American Society of Heating, Refrigerating and Air Conditioning Engineers
1791 Tullie Circle NE
Atlanta, GA 30329

The American Society of Mechanical Engineers
345 East 47th Street
New York, NY 10017

ASM International
Materials Park, OH 44073

The Institute of Electrical and Electronics Engineers, Inc.
345 East 47th Street
New York, NY 10017

Institute of Industrial Engineers, Inc.
25 Technology Park/Atlanta
Norcross, GA 30071

National Council of Engineering Examiners
Box 1686
Clemson, SC 29633

National Institute of Ceramic Engineers
757 Brooksedge Plaza Drive
Westerville, OH 43081

National Society of Professional Engineers
1420 King Street
Alexandria, VA 22314

Society of Automotive Engineers
400 Commonwealth Drive
Warrendale, PA 15096

Society of Manufacturing Engineers
One SME Drive
Box 930
Dearborn, MI 48121

Society of Naval Architects and Marine Engineers
601 Pavonia Avenue, Suite 400
Jersey City, NJ 07306

Society of Women Engineers
345 East 47th Street
New York, NY 10017

APPENDIX E

Books on Careers in Engineering

Beakley, George C., et al. *Careers in Engineering and Technology.* 4th ed. New York: Macmillan, 1987.

Collins, Steven et al. *The Professional Engineer in Society.* New York: Unipub, 1988.

Constance, J. D. *How to Become a Professional Engineer.* 4th rev. ed. New York: McGraw-Hill, 1987.

Crape, James R. *Engineering Career Package.* Canyon Country Calif: JRC, 1982.

____. *Power Plant Engineering Opportunities.* Canyon County, Calif: JRC, 1982.

Electrical Occupations: Equipment Planning Guides for Vocational and Technical Training and Education Programmes, Vol. 9. New York: Unipub.

Engineering Science and Computer Jobs. 9th ed. 1988.

From Tech Profressional to Corporate Manager. Pasadena, Calif.: T-C Pubns., 1984.

From Tech Professional to Entrepreneur. Pasadena, Calif.: T-C Pubns., 1986.

Gordon, Alan, et al. *Employer Sponsorship of Undergraduate Engineers.* Text ed. Gower Publishing, 1985.

Harmon, Margaret. *Ms. Engineer.* Louisville, Ky.: Westminster/John Knox, 1979.

Herden, Robert B. *Career Management for Engineers.* New York: Vantage, 1988.

Kamm, Lawrence J. *Successful Engineering, A Guide to Achieving Your Career Goals.* New York: McGraw-Hill, 1989.

Kaufman, Harold, ed. *Career Management: A Guide to Combating Obsolescence.* New York: Institute of Electrical and Electronics Engineers, 1975.

Kemper, John D. *Introduction to the Engineering Profession.* New York: Holt, Rinehart and Winston, Inc., 1985.

Leech, D. J., and B. T. Turner. *Engineering for Design for Profit.* New York: Halsted Printing, 1985.

Peterson's Engineering, Science and Computer Jobs. Princeton, N.J.:

Petersons Guides, 1989.
Pietta, Dan H. *The Engineering Profession: Its Heritage and Its Emerging Public Purpose.* Lanham, Md.: University Press of America, 1984.
Roadstrum, William H. *Being Successful as an Engineer.* San Jose, Calif.: Engineering Pr., Inc., 1978.
Robinson, Clark Z. *Top Dollars for Technical Scholars: A Guide to Engineering, Math, Computer Science and Science Scholarships.* Alexandria, Va.: Octameron Associates, 1986.
Schaub, James H., and Karl Pavlovic. *Engineering Professionalism and Ethics.* Melbourne, Fla.: Krieger, 1986.
Shackleton, S. Paul. *Opportunities in Electrical and Electronic Engineering.* Skokie, Ill.: National Textbook, 1982.
Shields, Charles. *Thinking about Engineering.* Princeton, N.J.: Petersons Guides, 1988.
Smith, Ralph J. *Engineering as a Career.* 3rd ed. New York: McGraw-Hill, 1969.
Smith, Ralph J., and Blaine Butler. *Engineering as a Career.* 4th rev. ed. New York: McGraw-Hill, 1983.
Sunar, D. G. *Getting Started as a Consulting Engineer (Engineering Career Advancement* set). Belmont, Calif.: Professional Publications, Inc., 1986.
Vetter, Betty, ed. *Opportunities in Science and Engineering.* 2d ed. Washington, D.C.: Commission on Professionals in Science and Technology, 1984.

INDEX